Memory

THE DARWIN COLLEGE LECTURES

Memory

EDITED BY

Patricia Fara *and* Karalyn Patterson

CAMBRIDGE
UNIVERSITY PRESS

CAMBRIDGE UNIVERSITY PRESS
Cambridge, New York, Melbourne, Madrid, Cape Town, Singapore, São Paulo

Cambridge University Press
The Edinburgh Building, Cambridge CB2 2RU, UK

Published in the United States of America by Cambridge University Press, New York

www.cambridge.org
Information on this title: www.cambridge.org/9780521572101

First published 1998
This digitally printed first paperback version 2006

A catalogue record for this publication is available from the British Library

Library of Congress Cataloguing in Publication data

Memory / edited by Patricia Fara and Karalyn Patterson.
     p.   cm. – (The Darwin College lectures)
Includes bibliographical references and index.
ISBN 0 521 57210 X (hardcover)
1. Memory. I. Fara, Patricia. II. Patterson, Karalyn. III. Series.
BF371.M4475 1998   153.1′2 – dc21   98-17394 CIP

ISBN-13 978-0-521-57210-1 hardback
ISBN-10 0-521-57210-X hardback

ISBN-13 978-0-521-03218-6 paperback
ISBN-10 0-521-03218-0 paperback

Contents

Introduction

PATRICIA FARA *and* KARALYN PATTERSON

In his *Vie d'Henri Brulard*, the French novelist Henri Stendhal articulated the difficulties faced by an autobiographer endeavouring to recapture his life of thirty years earlier: 'I make many discoveries . . . They are like great fragments of fresco on a wall, which, long forgotten, reappear suddenly, and by the side of these well-preserved fragments there are . . . great gaps where there's nothing to be seen but the bricks of the wall. The plaster on which the fresco had been painted has fallen and the fresco has gone forever.' The richness of human life depends on our ability to remember the past, yet – like Stendhal – we are painfully aware of our memory's selectivity and vulnerability. Whether we are trying to recall particular details of our own lives, or to construct historical narratives describing broader cultural changes, we must all confront the gaps and distortions inherent in recapturing the past. This perplexing faculty, so central to our existence, exerts a universal fascination: as individuals, we wish to learn more about how our own memory functions; as members of society, we are concerned to appreciate the multiple ways in which our history is preserved. What we remember is intimately linked to how we remember, but innumerable approaches have been devised to explore that complex web of connections. The eight chapters in this volume, all by leading experts in their fields, transcend this diversity to address together the relationships between individual experience and collective memory.

The word 'memory' reflects a richly layered palimpsest of connotations, but these can be grouped into two clusters of meanings. Memory may refer to our ability to remember the past, and so represent a function generally attributed to the brain. But it does, of course, also denote a more abstract concept of something – a person, a feeling, an episode –

which is itself remembered. Although they may initially seem distinct, these two facets of memory are closely intertwined. We expect scientists to be concerned with studying the processes involved in remembering, and humanities scholars to be interested in the products of memory, but this collection of essays exposes the falseness of such a dichotomy.

In his famous lecture of 1959, C. P. Snow influentially characterized the 'two cultures' divide of modern society into the sciences and the arts. This fundamental division stems from the work of Giambattista Vico, but was reinforced during the Enlightenment period, when scholars started to draw new disciplinary boundaries delineating the map of knowledge. The authors of the eighteenth-century French *Encyclopédie* split human understanding into three main areas: memory, which included the study of history; reason, comprising much of what we would call the sciences; and imagination, broadly speaking, artistic fields such as poetry and music. Although this fragmentation of knowledge has been perpetuated and strengthened during the past 200 years, it is now being challenged by an increasing number of academics. The approach of this book is intentionally interdisciplinary, seeking to explore the insights into memory which can be gained by juxtaposing the complementary perspectives of specialists who venture beyond normal disciplinary confines. As editors, we have arranged the papers in a continuous spectrum that extends, in conventional terms, from historical to scientific treatments of memory. However, they are bonded by several common recurrent themes, including memory's material embodiment, techniques of memorization, collective memories, reinterpretation of the past, and failures of memory. We hope that our readers will, like our contributors, come to appreciate how previous understanding can be enriched by reappraising a topic from fresh angles.

Mnemosyne – mother of the Greek Muses – was traditionally portrayed as a young woman dressed in green, crowned with ever-verdant juniper, and carrying her recording implements, a book and a pen. The dust jacket of this book shows how an important Enlightenment theorist of art, George Richardson, visually interpreted this symbolism which had been influentially codified 200 years earlier by Cesare Ripa. Although such allegorical representations now seem remote from our

2

understanding, they informed many European artists seeking to define moral issues in aesthetic terms.

Like Ripa and Richardson, we often perceive written texts to be the most vital repositories of our memories: the Solomon Islanders have aptly coined the pidgin expression 'House Belong Memory' to describe their national archives. But as the illustrations in this book demonstrate, memories are also enshrined in many other ways. For example, Australia commemorates its fallen soldiers in a cathedral-like Hall of Memory, where mourners wear rosemary, a floral symbol of remembrance since the Roman era. Around the world, museums preserve material evidence of former civilizations, while art galleries display pictures whose concealed symbolism reverberates with cultural memories. In addition, the patterns of ritual dances, religious ceremonies, songs and myths provide constantly changing reflections of older beliefs. Another of the aims of this collection is to explore the interconnections between some of these superficially diverse means of preserving memories.

Memory itself became an object of academic study during the nineteenth century, and the contributors to this book also examine some of the techniques which have been developed for investigating how memory operates. Explanations of mental faculties have ranged from the material to the spiritual. Although the goal of modern scientific techniques is to expose the brain's functioning down to the molecular level, long-standing debates about the nature of consciousness and the relationship between mind and matter remain unresolved. Just as cultural memories may be embedded in material artefacts, so too the memories of personal experiences leave permanent physical traces within our bodies. Whereas the phrenological belief (see Figure 1) that memory is located beneath a particular bump on the front of the skull now seems almost laughably naive, memories are, of course, neurologically coded by physiological and chemical processes, and certain areas of the brain are particularly significant for storing and retrieving memories. Over the centuries, people have formulated different analogies of brain function which reflect technological achievements of the period. Mnemosyne's book and pen recall the Roman view of memory as a wax tablet, or John

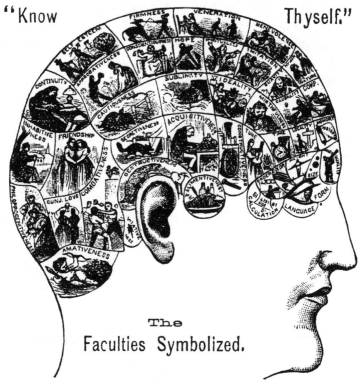

"Know   Thyself."

The
Faculties Symbolized.

PRINCIPLES OF PHRENOLOGY.

Figure 1 'The Faculties Symbolized', showing the principles of phrenology. Victorian phrenologists proposed various schemes for the brain's internal organization, relating specific faculties to the external structure of the skull that could be felt as bumps or depressions. In this example, Memory is located in the centre of the forehead.

Locke's *tabula rasa* (blank slate). In our century, the simile of a telephone exchange replaced older hydraulic and mechanical models, but has now itself been supplanted by increasingly sophisticated neural network models based on computers which have revolutionized our appreciation of learning processes.

In addition to these studies of the brain's physical structure, other routes towards a scientific understanding of memory have been developed. Psychologists of various schools have examined how we remember different types of information, and also analysed the impact of our remembered interpretations on our behaviour. But their conclusions are by no means restricted to scientific discourses. Particularly since the invention of psychoanalytical techniques, insights into how we remember, forget and interpret the past have been indelibly incorporated into our daily perceptions, thus permeating literature, art, and the writing of history. One of the strengths of the interdisciplinary approach of this volume is that by addressing shared issues from varied perspectives, its authors can explore vital connections between the physical and the psychological, the fictional and the factual, the material and the abstract, the individual and the cultural. These pairs are not opposed polarities, but reflect different facets of the complex entirety that constitutes memory.

In the opening essay, 'Disturbing Memories', *Richard Sennett* raises issues which recur throughout the collection. He argues that, although healing the wounds inflicted by negative experiences demands remembering these experiences collectively, the corporate economic structure of modern societies limits the therapeutic effectiveness of these shared memories. Drawing on his own case study, he describes how a group of fired IBM programmers in America constructed various retrospective accounts of their dismissal. Initially seeking to rationalize their unemployment by searching for vindictive premeditation on the part of their superiors, they subsequently attempted to lay blame on external agencies, such as increasing competition from lower-paid experts in other countries. In a distressing third stage, they assumed personal culpability. Modern capitalism, Sennett believes, encourages the feeling that memory is a private possession: whereas individuals desire to empower themselves through recreating their histories, large corporations are dissociating themselves from their institutional past.

Sennett is also interested in how the form of urban spaces is moulded by people's experiences and ways of living. In this chapter, he turns to the sociologist Maurice Halbwachs, who illustrated how successive rulers of Jerusalem rewrote the city's history to make it correspond to their own cultural views. This focus on the role played by material objects in reifying diverse historical visions is shared by *Catherine Hall*, in her essay ' "Turning a Blind Eye": Memories of Empire'. Like Sennett, she is concerned to explore how collective memories are created, since these processes can reveal many of the concealed suppositions structuring society. Hall conducts us around the Bournville theme park in Birmingham, built by Cadbury as a celebratory testament to the company's development from its Quaker origins at the end of the eighteenth century. She convincingly demonstrates how the conversion of chocolate into a national symbol of delicious good taste was embedded within the establishment of the British Empire. Through examining the iconography of the Bournville exhibits, she exposes some of the ways in which constructing a picture of former times inevitably entails suppression of events and attitudes. For Hall, this is not just a historical exercise: examining the fabrication of such collective memories is vital to enable the citizens of multi-ethnic Britain to understand – or re-remember – the racial tensions inherent in imperial growth.

Hall and Sennett forcibly remind us of the authorial influence in reconstructing the past, whether it be restructuring Jerusalem or embodying imperial aspirations in a heritage display. As scholars like Haydon White and Paul Ricoeur have emphasized, writing histories and writing stories are activities which resemble each other far more closely than many people would like to believe. In 'Memory and the Making of Fiction', *A. S. Byatt* provides a fascinating insight into the ways in which novelists weave into their books memories of personal experiences, both lived and read. As well as discussing the work of such diverse authors as Marcel Proust, William Wordsworth and George Eliot, she reveals, in a twinned display of erudition and imagination, how she herself uses mental markers to shape and memorize the narrative she is creating.

Just as we are no longer familiar with Mnemosyne's iconic attributes, so too we have forgotten ways of remembering. In Victorian England, soldiers were urged not to relax 'as you whiz along in a railway carriage', but to count the number of telegraph poles per mile in order to train their memorizing ability. Byatt discusses the classical art of training the memory by manipulating images in the mind, the 'art of memory' enthrallingly expounded by Frances Yates. Before printing was invented, these visual methods of association, which were to prove so influential on Western art and architecture, enabled many people to perform what seem to us quite prodigious feats of memory. Byatt relates Elizabethan memory theatres to writers' mnemonics, such as John

Milton's rhythmic lists, Proust's vision of a novel as a cathedral, and her own mental image of a dress knitted by the thread of language.

Mnemonic devices are also central to *Jack Goody*'s 'Memory and Oral Tradition', a rich analysis of recall in literate oral cultures designed to correct previous interpretations. Complementing Sennett's and Hall's discussions, this essay considers the nature of collective memories in oral cultures, and furnishes further examples of how memories can be embodied in material artefacts, such as the regal stools acting as visual prompts for narrators of Asante dynastic history. In a study of epic poetry ranging impressively from Homer to the Indian Rig-Veda, Goody concentrates on the Bagre compiled by the Ghanian LoDagaa people whom he has studied extensively in the course of his long anthropological career. Armed with a taperecorder, he has demonstrated that, contrary to earlier received views, long recitations are not learned by rote and repeated verbatim, but alter and develop as they are retold.

For Goody, these limitations of memory encourage the generative use of language which underpins cultural creativity and diversity. Like language, memory is a distinctive feature of human existence which enables us to learn and to produce meaning from the sensory data constantly bombarding our brains. As Stendhal observed in the quotation given at the beginning of this introduction, memory entails processes of remembering, but is also about forgetting. 'I've a grand memory for forgetting', quipped Robert Louis Stevenson; and a century ago, it was not memory but its absences which preoccupied Sigmund Freud. During the past decade, this topic has acquired a new urgency as debates have raged about the validity of recovered memories and – the flip side of the psychotherapeutic coin – false memories. *Juliet Mitchell*'s discussion of 'Memory and Psychoanalysis' is concerned with the operation and formation of unconscious memory, a complex subject which she treats by interweaving a nuanced account of American, French and British models of human behaviour with moving examples of patients' problems drawn from her own practice. Despite the current vituperative campaign colloquially labelled 'Freud-bashing', she demonstrates the depth of insight into repressed memories which a Freudian approach can bring.

In a sense, forgetting also forms the subject of *Barbara Wilson*'s contribution, 'When Memory Fails'. Like Mitchell's essay, Wilson's is enriched by studies of her own patients, but, in this case, people with brain disease or injury. Her compassionate discussion deals with pathological examples of the spectre haunting each one of us as we age – the prospect that our memories may fade, thus diminishing our humanity far more trenchantly than physical disability. After surveying different ways in which our brains store memories, and comparing various forms of amnesia and memory impairment, Wilson shows how people with severe yet specific areas of memory loss can appear to be functioning normally. More encouragingly, she presents various strategies for coping with memory problems incurred after illness or injury, ranging from straightforward tactics of labelling objects and compiling lists of tasks, to relying on the technological help afforded by computers. Wilson concludes with a poignant quotation from an apparently successfully rehabilitated young brain-injured man who commented that he no longer mourned the loss of his memory as he was unable to remember what it felt like to remember.

As the incidence of Alzheimer's disease escalates, many of us are concerned that one day we too may be unable to appreciate that we are losing this vital faculty of memory. At the end of 'How Brains Make Memories', *Steven Rose* stresses the value of scientific research by suggesting that his work may provide a new route for studying the progressive memory loss which characterizes Alzheimer's. Rose lucidly articulates the interdisciplinary fascination of memory, that intensely personal and subjective feature of existence which can, almost paradoxically, be explored by neuroscientists claiming to practise an objective research methodology. He describes how recently developed neuroscientific techniques are identifying the specific brain locations, neural activity patterns and biochemical processes involved in remembering, focusing in particular on his own experiments examining the learning processes of day-old chicks.

Terrence Sejnowski approaches the topic of learning from a different angle, that of computational neurobiology. In 'Memory and Neural Networks', he reviews the two broad classes of learning algorithms, supervised and unsupervised. During the 1980s, research focused on supervised learning, in which the desired output of the system is specified in advance, and the network learns by error correction based on this feedback. Sejnowski describes recent exciting work on networks which are not taught, but themselves extract information from natural sounds and images to generalize intelligently about the world around them. With its heavily technological bias, this study may seem remote from the historical and literary chapters with which this book opens. Yet the interrelationships bonding these apparently disparate approaches to memory are well illustrated by Byatt's comparison of neural networks with her own experience as a novelist of searching for connections, what she describes as 'trawling a network for flashes of significance, or preening hooks and eyes'.

'O Memory! thou fond deceiver', commented the English eighteenth-century writer Oliver Goldsmith in his ode to this elusive human characteristic. While marvelling at modern scientific advances, we still share Goldsmith's awareness of memory's deceptiveness, suggesting that it could forever evade the grasp of analysts seeking to define it. Memory may escape our understanding, but we hope that the essays in this collection will enhance our readers' appreciation of its eternal fascination.

1   Disturbing Memories

RICHARD SENNETT

Darwin's legacy

Charles Darwin is most famous as a natural historian, but his work had a profound influence on our understanding of memory because of his conception of biological time. Darwin set biology in a much longer context of time than had previous religious versions of creation and biological history, and so he created a puzzle: what is the relation of natural time, measured in millions of years, to the human historical time-frame in which we measure, say, the growth of classes or the development of cities in decades and centuries? Historical time seems a mere blip on the evolutionary scale. As the psychologist William James noted, our personal experiences of time are even more inconsequential in the Darwinian scheme of things, since personal events usually span mere days and months.

Moreover, Darwin gave natural time a distinctive character: he depicted it as conflictual and competitive. The concept of the survival of the fittest came late in Darwin's thinking, and he was ambivalent about the idea. He did not, like Alfred, Lord Tennyson, imagine nature 'red in tooth and claw', but he did think that what we call today an ecosystem depended on the ever-changing strengths and weaknesses of the species it contained, and that extinction and species failure were part of this natural order. Time is a destroyer as well as a creator.

What sense could students of human society and psychology make of time thus conceived? Most famously, for Herbert Spencer as for other social Darwinists, the survival of the fittest justified the pains of the capitalist order – but in a particular way. Natural time deprives human society of its idealism. Whatever might be contrived by ideas of justice contrary to the survival of the fittest human beings, Spencer argued, it is meaningless when set in the context of the *longue durée* of natural time. Capitalism is more 'natural' than socialism because its economic processes mirror natural selection in that *longue durée*.

But Darwin's work touched his social science contemporaries in a more humane way as well. To the great French neurologist Jean Martin Charcot, evolution seemed to explain why men and women are not at peace with themselves. As sentiments multiply and grow various, Charcot said, they conflict and compete with one another in the human heart; men and women are as much at war within themselves subjectively as they are at war with one another in society – subjective warfare is 'natural' and the self we become is like a species which has survived these inner battles.

It is this latter, more humane, response to Darwin's discoveries that has shaped the way we now think about memory. Memory is consciousness of time which has passed. In the Darwinian time-frame, our memories are trivial records; we cannot, obviously, summon to consciousness the millions of years which formed us as a species. But the conflictual, competitive Darwinian order of time obviously troubles also those acts of recovered time in which we are able to engage. Rather than a halcyon recovery of a golden age, of an Edenic pastoral, a more accurate act of memory reveals to us an unending record of pains and struggles which has formed us into the creatures we now are. Truthful memory opens wounds which forgetting cannot heal; the traces of conflict, failure and disaster are never erasable in time. The science of evolutionary reconstruction seemed to offer testimony to those ineradicable traces. There is no solace in the truths of memory.

Karl Marx came independently to this view in his studies of the revolutions of 1848. He detected revolutionary nostalgia, evinced by insurgents that year for the Great Revolution of 1789, and his later reading of Darwin strengthened him in his hatred of nostalgia – which is, Marx said, memory's lie. Sigmund Freud, who as a young man was an avid reader of Darwin, learned from Charcot that the gates of memory shut when the inner warfare described by Charcot becomes unbearable; truthful memory of the past requires courage as nostalgia does not, since the collage of conscious pain only becomes thicker when we remember well.

Yet if *what* we remember offers no solace, *how* we remember these wounds seemed to many of Darwin's heirs to offer some release. Like Freud, Marcel Proust came to that view of the work of memory in the final volume of *À la Recherche du Temps Perdu*; remembered pain or conflict can be objectified as present pain cannot be. In my own field of sociology, a similar emphasis on how we remember has shaped the study of collective memory.

That study began with Émile Durkheim. When families or communities share their memories, he argued, they draw closer together; they create a sense of solidarity through remembering together, even if what they remember is traumatic, like a war or an economic disaster. That is, collective memory is a source of social strength. Durkheim knew enough to recognize that these collectively shared memories might be anything but objective and factual; the degree of group solidarity might instead depend on their mythic properties. Yet he continued to hope that, by sharing rather than privatizing memories, communities might find a way to tell the truth about themselves. The question is, how? For Durkheim, the more varied the voices engaged in that common dialogue, the more likely the accuracy of what would be recalled: he is the first liberal theorist of memory.

I have been thinking about Darwin's age and legacy in trying to understand a set of social problems which may seem unrelated: the ways in which people today make sense of a new chapter in the history of capitalism, a chapter creating a flexible economy and undoing the protections and supports of the welfare state. Certainly social Darwinism of a sort has reappeared in this new order, which emphasizes the survival of the fittest in terms of those with technical skills and flexible (i.e. adaptive) behaviour patterns in their work. What has struck me is the difficulty people have in making an accurate record of what has happened to them in the course of this political-economic change: the denials of injuries sustained in the course of recent experience, the tendency of communally shared memories to generate myth rather than accurate histories.

I believe that people can indeed deal with the wounds of memory in their life histories only by remembering well (that is, by becoming more objective) and I am convinced that Durkheim's liberal faith remains compelling. Remembering well a shared injury is something which people cannot do by themselves, but must be shared by a group of diverse voices. However, the organization of the modern economy seems to work against people remembering well together; it does not support them in their efforts. And the reasons for this, I am trying to understand.

Recall and reconstruction

To make the difficulties of collective memory a little clearer requires an elaboration of a distinction which psychologists used when they first began to explore systematically the wounds of memory. On this subject, in the late nineteenth century, they were divided into two camps.

Some, mostly German, focused on the act of recall; that is, the circumstances under which we can accurately summon facts or events from the past. The psychologist von Sieberling, for instance, showed people pictures of the Kaiser at Sans Souci, and asked them a day later to recall as much as they could about the details of the photograph. Another school of psychologists, led by the American William James, focused on memory as a process of reconstruction. This school explored the storytelling by which we fashion a verbal account of what happened in the past.

These two aspects of memory intertwine. Our powers of recall can caution or rebuke the stories we make of the past – as when we recall an inconvenient fact about a parent which does not fit into the place we have assigned that hapless creature in the heroic narrative of our own growing up. The stories through which we shape memory in turn prompt us to recall some images or facts, while consigning others to the neural dustbin. Indeed, we seem quickly to forget particular facts if these are not fitted into some relational context; narrative sequences provide a far more powerful sense of context than do static images like a photograph.

In the early days of the psychological investigation of memory, neither school of research concentrated on disturbing memories, those memories which might inhibit recall of particulars or suppress stories too unsettling to recount. That task fell to Freud. He discovered traumatic events to be the source of repression, and repression to be expressed through a blanking out of memory – I use the word 'discovered' because, though these have become familiar ideas to us, they were not so a century ago.

Freud's work on repression intersects with James' on narrative within the world of dreams. Freud explored how the act of dreaming set people free to create incidents and forge connections between events which they could not do in a waking state. The dream is our unobstructed terrain of narrative. But the dream becomes a therapeutic narrative only when written or told in such a way that it prods the waking memory to name or specify something in response. Reconstruction then leads to recall; the dream work becomes objective.

Freud's idea of objectification thus rests on creating a *correspondence* between dream stories and lived experience. Many modern psychoanalysts have kept the emphasis on narrative construction but dropped the belief in correspondence. They encourage a person to reconstruct alternative, equally plausible narratives about the past, believing that, just as a novelist explores different versions of a story, so can traumatized patients. Again, as in Freud's therapies of recall, the disturbing events are to be contained and tamed within the play of language, but now the emphasis is on freely telling, rather than naming and connecting.

If you are a good Durkheimian, as I am, you will opt for narrative as an instrument for provoking recall, as Freud himself did. But more, you will believe that correspondence between recalling what actually happened – I use this old-fashioned term without shame – and the story of what happened depends not on a single narrator, but rather on a plurality of contending voices speaking to one another. Put formally, the interaction of recall and reconstruction occurs only under conditions of diversity: many contending narratives are necessary to establish painful social facts. It is only the noise of contention which wrests collective memory from that shared, dream-like state we call myth-making. Contending narratives need not resolve into one story in the process of searching for what actually happened – and this is where the modern Durkheimian parts company with social thinkers such as Jürgen Habermas, who believes in a gradual convergence of accounts into a single, shared story. But although the process of recall is shared, this elucidation re-creates incidents and elements that can be made into contending stories.

This brief foray into theory is necessary to understand what occurs in the life histories with which I have been concerned in my own work:

people who have shared common injuries at the hands of the modern economy and are seeking to interpret what happened to them. There has been a breakdown in the process of establishing correspondences between their narratives and events which actually occurred: they have become free, mythologizing narrators.

Remembering together

I am going to give as an example the operations of recall and reconstruction, naming and telling, in the lives of a specific group of middle-aged, middle-class American adults grappling with an upheaval in the modern economy. Many of them have been made redundant at a too early age, others face the threat of unemployment. They grew up in a society and in a generation which promised them that their working lives would take a more stable and rewarding course. The specific case that I will describe concerns computer programmers who have worked for the IBM corporation at an office near where I live in the country outside New York.

The events which have disturbed their lives are quite specific; starting a decade ago, they began to be dismissed as the corporation changed course to focus on personal computers rather than the large main-frame machines on which these programmers had worked. The redundancies at IBM intensified six years ago, as the corporation made use of equally competent engineers in India, who write computer programs for IBM at a seventh of the wages the Americans had received, programs which – thanks to the wizardry of the Internet – are transmitted to the home office in a few seconds.

When the American programmers began their professional lives, IBM promised them lifetime employment, generous benefits extending even to a corporate golf course, and a clearly outlined map of future possible promotions. It was the corporation as father, resembling General Motors and AT&T, as well as large Japanese corporations. Those days are no more. IBM has downsized, eliminated administrative layers, and made its corporate map more fluid in the course of reorganizing its business focus. Unemployment has thus become a middle-class, rather than just a working-class, phenomenon. The survivors of each wave of cuts live in fear about where the axe will fall next.

Some of these unemployed men spend time at the Coach Inn, not far from their old offices. It is a cheery hamburger joint formerly tenanted during daylight hours by women out shopping or sullen adolescents wasting time after school. It is here that I have heard these white-shirted, dark-tied men, who nurse cups of coffee while sitting attentively as if at a business meeting, attempt to make sense of their histories. Their collective act of memory has gone through three stages so far.

The first stage involved dredging up corporate events or behaviour in the past that seemed to portend the changes which subsequently came to pass. These acts of recall included such bits of evidence as a particular programmer being denied use of the golf course for a full eighteen rounds, or unexplained trips by the chief programmer to unnamed destinations. At this stage what the men wanted was evidence of premeditation on the part of their superiors, evidence which would then justify their own sense of betrayal; the story was at first that 'they', the authorities, knew all along what was coming.

But this collective narrative would not wash. For one thing, many of the superiors who fired them in the early phases of corporate restructuring were themselves fired in later phases, and now can also be found at the Coach Inn. Again, the company was in fact doing badly through much of the 1980s and early 1990s, unpalatable facts all too amply recorded in its annual balance sheet; the terms of the old corporate culture were plainly at odds with the health of the corporation. Finally, as reasonable adults these programmers came to understand that a history of premeditated betrayal converted those who had injured them into stick figures of evil, and made this ailing albatross of an organization into a more potent force in fantasy than it was in fact.

So in a second stage of rewriting their collective history, the IBM programmers focused on the Indian programmers and on the Internet, seeking external agencies to blame. The belief that foreigners undermine the hard-working efforts of native Americans is deeply rooted in American society, a prejudice that extends back to native reactions to southern and eastern European immigrants in the nineteenth century. Other countries may share a similar view of foreigners, but few others have been as necessarily open to immigration in the past as the resource-rich but population-deficient land of America. Now the programmers in

India were shoved onto the stage of this old fear, and the Internet served as their Ellis Island, their port of immigration. For a few months, the men at the Coach Inn sought to explain their own troubles as due to 'outsiders' taking over the corporation (the fact that the new president of IBM was Jewish was noted), and in turn that view led them, in 1994, to vote for extreme right-wing candidates whom they would have found absurd in more secure times.

Yet again this shared story would not wash. Part of the reason was that it conflicted with their belief, as scientist engineers, in the virtues of technological developments like the Internet. They also acknowledged the quality of the work coming out of India. Most of all, this narrative version was thin. It referred only to the actions of others, unknown and unseen elsewhere; the story-tellers became passive agents. During the stage in which the programmers constructed the perfidy of the Indian wage-busters and the machinations of IBM's Jewish president, they had little to share with one another about their own work experience, excluding much personal exchange. And so gradually the programmers put it aside.

The third stage of constructing a collective history has rectified this sense of passive agency, but in a distressing way. Gradually the discussions at the Coach Inn have evolved into considerations of what the unemployed men could and should have done earlier in their own lives in order to prevent their present economic condition. These discussions are focused on the fact that IBM itself stayed committed to main-frame computers at a time when growth in the industry occurred in the personal computer sector; most of the programmers, as I noted, were main-frame men. In the third stage of remembering they recited the successes of people who ten or twelve years previously had moved into the personal computer sector, via risky small businesses; this is what the programmers at the Coach Inn think they should have done. In this, they blamed themselves for having been too security dependent, for having believed in the promises of corporate culture, for having played out a career scenario not of their own creation. The history of IBM's blunders, the advent of the Indian programmers, the corporate downsizing motivated by investors who wanted short-term high returns on their stock-holdings – all are now subsumed in a more personal sense of lost opportunity for which the programmers felt personally responsible. Here, to date, the shared story has rested.

As a friendly neighbour and nostalgic Marxist, I wanted the programmers to hold the corporation instead of themselves accountable for their difficulties; I wanted them to organize and foment rebellion among the survivors; I wanted them to act on their memories. But then, I am a tenured professor at a rich university; as the coffee gatherings at the Coach Inn evolved more personally, I fell increasingly silent. This is not Marx's territory, more the terrain of Durkheim's fear, but with a twist. The men have come to feel close to one another in reminiscing, but they have become more individuated and inward-turned in relation to the larger society. They jointly speak with a single voice, but it is the voice of 'I' rather than 'we'.

We can observe a changing balance between recall and reconstruction in the evolution of these memories. The free play of reconstruction operated throughout, but the work of recall suddenly diminished. The first two versions of dismissal were checked by objective events or historical conditions: the first by the men's direct knowledge of IBM's condition, the second by the belief in technological progress which lies at the heart of their sense of personal integrity. The third version, however, dispels the bite and power of those facts. Instead this version makes the past a projection of the programmers themselves. The story of personal failure puts them at the very centre of the action as narrators.

Why did this happen? One reason was that sharing the past in this third way legitimated open, personal discussion between the people in the Coach Inn; the discussion came to be tinged with an air of confessional intimacy. This is the most interactive, the most sociable version of their history. Again, the third version relieved them of struggling any more – a respite many middle-aged people wish for. This is perhaps why the programmers spoke more with an air of resigned finality than anger about being 'past it', about having blown their chances, even though they are still physically strong and a long, uncertain life stretches before them.

Within the flow of memory itself, this third version gave the narrators something more, which could be called subjective solidarity. All the facts and impersonal events connect to the question of their own will and choice; the third version is a story with a solid centre – me – with a well-made plot: 'what I should have done was take my life into my own hands, like the programmers who started small new companies'.

It may go against common sense to think of memories of failure as solid and centred, and I have no wish to deny the pain these men are feeling. But they have, in the flow of memory, at last possessed their past in a stable form; memory has come to rest; their sense of present and future time has become static.

Of course all specific cases are special cases, but I think this particular one is more generally suggestive. The cautionary, hard work of recall is active when narrative is unstable, but fades as the narrative crystallizes and comes to rest. The sense of being at the centre of a memory, the memory's subjective solidity, serves social solidarity as well. And unfortunately, in reconstructing the past, relief proves no stranger to resignation.

This episode shows concretely what happens when remembering together gradually resolves itself into a single voice, when collective memory becomes a group's single 'I' speaking. This was not what Durkheim hoped for; one of his students has illustrated the alternative in which he placed his hopes.

The restlessness of memory

The sociologist Maurice Halbwachs was a follower of the sociologist Émile Durkheim, a professor in Strasbourg and in Paris who led a quiet life until arrested by the Nazi authorities. He was sent to Buchenwald and murdered there for asking the police what had happened to his wife's Jewish father. He died, it is not too much to say, for the sake of an accurate account of the past. Halbwachs' writings now strike many readers as flaccid; too often a dutiful student, he sought to clarify details in Durkheim's theorizing. But Halbwachs came into his own in writing about collective memory; he brought the liberal theory of memory concretely to life.

He did so in a study of how the various occupiers of Jerusalem over the long course of its history – Jews, Romans, Christians, Muslims – each had to rewrite the history of the city when they occupied it. He noted how each ruling power had to shift the centre of Jerusalem, in order to make the design of the city accord with their own view of themselves and history. For early Christians, for instance, the Holy Sepulchre was the city's centre, because the Second Coming seemed imminent; when the Crusaders reconquered the city, they renewed the route of the Via Dolorosa, along which occur the Stations of the Cross, since the street echoed the Crusaders' own experience of an arduous pilgrimage filled with suffering (Figure 1). The same shifts marked the city when it was in the hands of other faiths. For the Muslims in the Renaissance, the Western or Wailing Wall, sacred to Jews, was merely the physical foundation for the Dome of the Rock, paralleling the Islamic belief in Judaism as a preparation for the appearance of the Prophet (Figure 2).

Halbwachs wrote as a Christian, and his urgent purpose in this study of the shifting centres of Jerusalem was to impress upon other Christians that there is no durable spiritual centre in the world. Sociologically, though, Halbwachs argued that only when Jews, Christians, and Muslims understand that theirs is not 'the Jerusalem'; only when they acknowledge that each faith has shifted its own sacred locus in time as the city evolved; only then could these religions become more mutually tolerant, because they would be less anchored in their own moral geography.

Halbwachs had specific ideas about how the work of remembering might lead to this sort of social relationship. Recall will remain active only if narrating remains restless. The facts of the past have to be used to combat people's tendency, in his words, to 'centre themselves in their memories', but one cannot suppose that people will be their own severest critics. For Halbwachs, the critics can only be people who have a different 'centre' – who have differing stories of the same wars, strikes, or the meaning of particular places. These critical voices both stimulate recall and destabilize narrative reconstruction. In literature terms, a searching memory requires a de-centred subject.

The sociological problem of collective memory is how to compose a body of critical voices that will de-centre one another. From Halbwachs' perspective, the computer programmers made a fatal error in choosing the Coach Inn as a place to meet; it was too intimate and they were too alike. They should somehow have managed to hold their discussions on

Figure 1 Jerusalem's Via Dolorosa, showing the arch known as 'Ecce Homo' (the second Station of the Cross).

Figure 2 In Jerusalem, the Dome of the Islamic Haram al-Sharif is bounded by the Western (or Wailing) Wall, a vitally important Jewish religious site.

IBM's turf, and included in the group their former bosses, and perhaps even the Indians via the Internet. Under these adverse circumstances, a particular account of dismay would not be likely to go unchallenged, and so would be less likely to come to rest.

At least this is what Jerusalem, which he never visited, suggested to Halbwachs, who wrote at a time when Europe lay in the grip of punitive and violent myth memories. Remembering well requires reopening wounds in a particular way, one which people cannot do by themselves; remembering well requires a social structure in which people can address others across the boundaries of difference. This is the liberal hope for collective memory.

This may seem highly idealistic; common sense, that most fallible of guides, tells us, for instance, that people shy away from conflict. But in fact people dwell in this state all the time. Not only business competitors but lawyers, diplomats, and labour negotiators live off debating contrary narratives and are stimulated by it. The sociologist Lewis Coser long ago showed how verbal conflict in fact creates social bonds between dispu-tants. Arguing supposes you care what the other side thinks, and the verbal process itself – its feints, shifts of argument, offers of compromise – gradually knits people together.

So the problem is why, in the kinds of economic experience of unemployment and premature uselessness I have described, those conflictual relations are not taking form, why collective memory of shared injury can become a detour rather than a confrontation with capitalism's current pains.

Memory as a personal possession

There are a host of institutional reasons for this detour. The modern economic system is intensely individualizing; as it strips away collective supports and defences against economic dislocation, it forces people to behave as isolated entrepreneurs, even when they work in large organizations. The very character of work has become redefined as a task unrelated to past performance or to future embedding within a bureaucratic structure. An executive for AT&T recently summed up how the flexible economy is transforming work: 'In AT&T we have to promote the whole concept of the work force being contingent, though most of the contingent workers are inside our walls. "Jobs" are being replaced by "projects" and "fields of work".' The old concept of a career, moving from well-marked stage to well-marked stage in the course of a lifetime, is dead. People will no longer spend the whole of their working lives in one company. Instead, the reality now facing young American workers with at least two years of college education is that they will change jobs on average at least eleven times in the course of their working lives, and the skills they need to work will change at least three times. All this means that the narrative of a life is no longer a tight, well-made story.

Institutional, individualizing insecurity makes subjective life itself highly unstable and uncertain. Shoring up the sense of self in a world without reliable institutions thus becomes urgent psychological business, as the economy deforms the coherence and continuity of lived time. And here, I think, is where the work of memory enters.

The final stage of memory among the unemployed computer programmers yielded them a sense of subjective solidarity. I want to argue that, by extension, under current conditions the work of memory, much

Richard Sennett

more than momentary sensation or the operations of deductive reasoning, holds out the promise of making inner life feel definite and solid. It
is through memory that we seem to possess ourselves. Put another way,
memory becomes like a form of private property, to be protected from
challenge and conflict – a property of the self which those exchanges
might erode.

Modern capitalism encourages such feelings about memory as private
property. In their emerging forms of growth, chameleon institutions
such as IBM are undoing their own institutional memories; the businesses want to change rapidly, and fear being stuck in the past. They
thus shy away from rewarding length of service or experience. De-
bureaucratized institutions instead reward people not prone to institutional attachments; in the computer business, for example, employers
now put a greater premium on hiring those who have moved from place
to place than upon those who have remained steadily at one firm. Institutional loyalty is seen as a portent of personal dependency, a lack of
enterprise; people with long institutional memories are a drag on
change.

Under these conditions, a negative value is placed on the work of
recall. Unlike the bureaucratic dinosaurs which rigidly graded length of
service and operated by seniority, a new-style corporation has no interest in its own history, save as a caution about what needs to change. It
certainly has no interest in provoking its employees to remember; it is
not an interlocutor of time. And for this reason, memory has become
increasingly subjective and internal, the *longue durée* of personal private
experience.

Given these realities, it becomes indeed reasonable to see the domain
of memory instead as a private matter, an archive one wants to protect
from the violations of the competitive world. People will express their
memories in a way which empowers and solidifies their remembering;
by sharing those reminiscences they will create bonds of trust and loyalty
missing in the corporate world. I would never say of the dismissed computer programmers I know that they have acted irrationally; rather, self-
destructively. A business magazine recently reported the 45 000
employees who would be dismissed by the AT&T corporation in 1996
were taking their imminent redundancy with resignation rather than

anger. Shared memories do not protect the workers I have interviewed against the world. They acknowledge this in an inverted way, by speaking of themselves as 'past it', by focusing on what they should have done when there still was time.

My claim is that people do not remember well because the modern economy does not encourage it. Defence lies not in repression of the past, but in narratives which empower the remembering subject, and this private empowerment is an act in which people can collectively collude. Durkheim's and Halbwachs' liberal dream has been defeated by the denials of time in the economy and, perversely, by the freedom of the narrating subject.

Conclusion

This, at least as I understand it, is a large part of the crisis of liberal culture in modern capitalism. It is a complicated story in which people's sense of time is victimized both by an economy with no interest in duration, stability, or continuity, and by the way people construct these values of time for themselves.

In a way we have returned to the problem Darwin posed to his contemporaries: how can we make the time that we have lived matter in the conduct of our lives? In Darwin's era, the length of biological time seemed to make experienced time insignificant; in our era, the shortness of economic time seems to result in the same sense of insignificance.

One of the oldest meanings of the word 'suffering' is to feel a passive victim of one's fate. The middle-aged workers I have interviewed are struggling against that sense of passivity, but they are not finding in the labours of memory a remedy against it. The cultural problem of time in modern capitalism requires something like the conditions of place Halbwachs imagined when thinking about Jerusalem – a confrontation between voices in conflict, a space in which the wounds of lived time can be effectively voiced. The subtlety in this process, as I have tried to make clear, is that the victims can develop strong voices only when they become de-centred, only when – through the restlessness of how they recount the past – they search for some correspondence to what actually happened to them. This, as Freud knew, is painful work for memory coping with reality.

Richard Sennett

FURTHER READING

Connerton, P., *How Societies Remember*, Cambridge: Cambridge University Press, 1989.

Halbwachs, M., *La Mémoire Collective*, Paris: 1950.

Healy, C., *From the Ruins of Colonialism: History as Social Memory*, Cambridge: Cambridge University Press, 1997.

Kroyanker, D., *Jerusalem Architecture*, London: Tauris Parke Books, 1994.

Nora, P., *Les Lieux de la Mémoire*, 3 vols. Paris: Gallimard, 1992. [English translation: *Realms of Memory, Rethinking the French Past*, transl. by A. Goldhammer, vol. 1, New York: Columbia University Press, 1996.]

Samuel, R. and Thompson, P., *The Myths We Live By*, London and New York: Routledge, 1990.

Sennett, R., *Flesh and Stone: The Body and the City in Western Civilization*, London and Boston: Faber and Faber, 1994.

2   'Turning a Blind Eye': Memories of Empire

CATHERINE HALL

In April 1991 'Cadbury World' was opened by the then Prime Minister, John Major, in the multiracial city of Birmingham, England. 'Cadbury World' is a living museum, the museum of Cadbury, the celebrated Quaker family who established a world famous cocoa and chocolate business and made their name a household name. 'Cadbury World' replaced the factory tours and educational visits which Cadbury provided until 1970 on the Bournville site. After the factory was closed to visitors the firm received thousands of requests for information and finally decided to construct a purpose-built 'experience' – the chocolate experience – next to the factory. This chocolate theme park attracts half a million visitors a year. It combines an exhibition with a brief factory tour. It aims, in the words of the accompanying brochure, to 'reflect the heritage of the UK's number one chocolate company'.

It tells the story of Cadbury, the Quaker family firm which began in the 1830s as a small retail business in Bull Street, Birmingham, went through difficult times in the 1850s but by dint of the hard work of two Cadbury brothers was successful enough in the 1870s to warrant the purchase of extensive land in Bournville and build the 'factory in the garden', which became famous the world over. There the large modern factory was combined with a model village and community facilities to provide a whole way of life for Cadbury employees. Between 1879 and 1931 Cadbury grew from a small family firm to the twenty-fourth largest manufacturing company in Britain. Bournville employed 300 people in 1879; this had become 2500 by 1899 and 14 000 by the 1950s. This work force has now substantially shrunk with the onset of automation. By 1990 the firm had one-third of the UK confectionery market. The family firm became a private limited company and, eventually, Cadbury Schweppes.

Catherine Hall

I have chosen 'Cadbury World' to tell a story about imperial history and to raise some questions about history and memory. The device of the case study is used to provide one example of a process which I am suggesting is pervasive: part of British national culture. There is nothing peculiar to Cadbury in the argument which I am making: rather, the firm's willingness to tell its history makes possible an analysis which could be widely applied, to firms, to museums, to the cultural forms which came out of empire. I aim to suggest some possible issues as to how history stories are currently being told and evoke some new ways of telling.

British national identity

The current debate over national identity in Britain is usually posed in terms of Europe – what kind of nation can 'we' be now that 'we' are losing significant elements of national sovereignty to Brussels? How is Britain to coexist with the rest of Europe? What will happen to our unique heritage and sense of distinctiveness? Another way of posing these questions might be to ask, what kind of a nation is contemporary Britain now that it is no longer at the heart of an empire? English and British identities, it could be argued, have been held relatively secure for centuries by notions of empire. England has been an imperial power for a very long time, the nation's sense of worth established in part by dominance over neighbouring Wales and Scotland and the colonization of Ireland, ever extending through conquest and settlement across the globe (North America, islands in the Caribbean, India, Australasia, parts of Africa and elsewhere). English identities have thus been historically rooted in notions of England, and white superiority over those other people and places which made up the British Empire. The break-up of that empire, the long process of decolonization which is still not fully completed, thus raises problematic issues about national identities. For if those identities have in part been held secure by certainties as to the relative place of centre and peripheries then what happens when the global map has shifted so decisively? England has been a colonizing power since she became a nation. Not surprisingly, therefore, nation and empire, national identities and imperial identities, have been very closely interconnected and nation and national identities have to be rethought when empire has gone. Who are 'we' when 'we' are no longer the 'master race'? Who are 'we' when 6.3% of the population are now classified as 'ethnic minorities' and in urban areas such as London and Birmingham that proportion is over 20%? Who are 'we' when significant numbers of decolonized peoples from Jamaica, Trinidad, Barbados, Guyana, India, Pakistan, Bangladesh, Kenya, Uganda and many other erstwhile British colonies have made their homes in Britain together with their children and their children's children and constitute an important part of that 'we'? Who are 'we' when many of 'us' are black or brown?

This question about national identity is a vital political question and one which has proved uncomfortable for many politicians. The right wing of the Conservative party focuses on 'our island story', hoping to save the nation by retreating from Europe into the fastnesses of England and insisting that the black population can be integrated only if they become culturally English or British. Others look to Europe and see the 'region' as the source of new identities for the twenty-first century. A part of this turn to Europe is the discarding of uncomfortable memories of empire, a process in which other erstwhile European imperial powers are also engaged, for the Dutch, the Spanish and the French have as much to forget as the British. A substantially white 'Fortress Europe', protected by its common policies on migration and asylum, can hope to hold at bay those economically poor peoples many of whom come from places once colonized by Europeans, and who now long to join the rich countries of the North.

What other possibilities might there be to construct new identities? Other places, for example Australia, might offer us some clues. Australian nationhood was sealed in the creation of the Federation in 1901. The nation state which was legitimated at that time was imagined as white, the 'White Australia' policies on Aboriginal citizenship securing the apparent erasure of those peoples and white immigration policies designed to secure the racial purity of the nation. In the current debates on Australian national identity, fuelled in part by the centenary of Federation and the possibility of a republic, some have argued for the conception of a 'post-nation', a nation which would no longer hold to the principal meanings of nation associated with the moment of 1901. Such a vision would entail discarding the notion of a homogeneous nation state with singular forms of belonging, making a new settlement with Aboriginal peoples, and imagining a different kind of future, one that depends on notions of mix and refuses ethnic purity. An analogous vision for Britain would mean reimagining the past, necessitating the abandonment of the link between nation and ethnicity ('we' are no longer a fictionalized 'Anglo-Saxon' nation), and indeed between nation and skin colour, and envisaging a new kind of nation which is not ethnically pure but rather inclusive and culturally diverse.

Historians and the histories which they have written have always been central to the construction of national identities. History emerged as a discipline at the same time that 'the nation' and forms of nationalism became central to Europeans. The many versions of 'our island story' have told us who 'we' are and how we are placed in the world.

In the mid 1990s history is probably as popular as it has ever been in Britain, not in its academic form but in its popular manifestations, on television, in the cinema, through the heritage industry. Genealogy and other forms of family history have made local record offices, once quiet spots, into bustling centres of activity, drawing in sections of the community who once would have been astonished by the idea of an encounter with the census or a rate book. Film and television have both catered for and encouraged the fascination with the past. The lively debate on the status of history in the last decade has marked the recognition of the importance of history in the making of the nation. The ascendancy of the Right through the 1980s meant that culture was enlisted to construct a particular notion of Englishness associated with heritage, conservation and the thematization of Englishness. This has been criticized by some writers for its nostalgic recreation of a rural England untouched by capitalist relations or exploitation. Others have welcomed unreservedly the new forms of popular history for the ways in which they have allowed ordinary men and women to articulate their interests in their pasts. Meanwhile the debates on history in the national curriculum marked one of the sites where the different possible versions of English/British history were contested by diverse groups – a sign of the significance of these issues beyond the schoolroom.

Re-remembering the British Empire

In her novel *Beloved*, Tony Morrison has argued for the 'memory work' which needs to be done so that African-Americans and white Americans can recover the history of slavery and understand its foundational place in the construction of the United States. That past has been too painful to remember but Morrison suggests that, if it is not re-remembered, it will haunt and disrupt contemporary society. In *Beloved*, the ghosts can only be settled when re-remembering is achieved. Only through the work of 're-memory' and new forms of understanding can that history be reworked in such a way as to make it liveable for the present, and make the present fully lived. Each generation has the task of making new histories, making a space in the mind for the past, a past reconstituted in the present.

As an historian in 1998 I want to suggest that unless the legacy of the British Empire is re-remembered it will continue to disrupt and unsettle our present, in ways that obstruct the development of a new kind of nation. The legacy of that empire is all around us and yet there is great reluctance to think about it or acknowledge its place in our history. Nostalgia envelops some aspects of it; commodification has provided another way of dealing with it, offering us packaged mementoes of an imperial past; yet neither helps us to assess the significance of that legacy in our present.

Memory, as we have learned from a variety of sources, from Sigmund Freud and the psychoanalytical tradition to the work of oral historians, is an active process which always involves a process of forgetting as well as remembering. Screen memories, as Freud named them – that is memories which may be composites of real events and fantasy – may enable us to forget very effectively. What we remember and how we remember it is crucial. Work on individual memory can offer a starting point. Drawing directly on the story of Oedipus, the analyst John Steiner explores the phenomenon of what he calls 'turning a blind eye', which refers to the process whereby an individual has access to forms of knowledge but chooses consciously or unconsciously to ignore it. Steiner argues that Sophocles simultaneously allows two interpretations of the play. The first presents Oedipus as the victim of fate and the dramatic sequence concerns his discovery of who he is, what he has done and to whom. The second is associated with 'the blind eye'. All the key participants in the story of Oedipus, he suggests (Oedipus himself, Jocasta and Creon), knew the truth at some level before the full weight of the knowledge was accepted by them. They colluded in the choice to ignore and evade their knowledge. Those evasions can lead to an amalgam of manoeuvres designed to deny and conceal. The cover-up evades the recognition of the crime, the facing of guilt and consequent possible reparation. People turn the 'blind eye' when seduced by ambition, as Oedipus was, and assisted by chance. Those around collude in this process and enjoy the benefits of not seeing. 'Turning a blind eye', Steiner argues, 'is an important mechanism which leads to a misrepresentation and distortion of psychic reality'. Such an interpretation of the play and the meaning it can offer does not invalidate the familiar reading of the play in terms of fate: in this account the focus is on Oedipus as the victim of fate who bravely pursues the truth. This was the reading which Freud drew on when he made the analogy with the ways in which the structures of the unconscious are revealed during analysis. Rather it can deepen our understanding of individual psychic processes and perhaps of collective ones too. Steiner recognizes the social and political potential of his argument. People turn a blind eye, he points out, to major dangers in our contemporary world, even when all the information pointing to the fragility of the situation is available.

I want to develop the analogy which Steiner makes between individual and collective processes and construct an argument about collective memory. My starting point is that Britain's imperial history is a deeply disturbing history that has involved expropriation and exploitation, relations of domination and subordination, and multiple racisms. That past, I suggest, haunts the present and prevents us from imagining a new kind of future. We, all those of us who live in Britain, need to work to remember that past differently, to recover those aspects which have been forgotten and evaded. Challenging versions of the past that are available, engaging in conflict over the existing representations of imperial and colonial histories will facilitate thinking about the present differently. Those versions of the present which say 'send "them" home', or 'chuck "them" out' or symbolically or socially exclude 'them' because 'they' are seen as the problem will only provide short-term answers, for repression comes back to haunt. Those named as 'outside' are always 'inside' as well. 'We' are discursively constituted through the binary negative of 'you', but that exclusion is critical to identity. Labelling or marginalizing black people cannot be a way of coming to terms with the interrelated histories of the peoples of the Empire, locked in their differentiated power relations. Rather we need to grasp the ways in which the English and their 'others' have been implicated in colonial relations; the ways in which slavery, for example, has been an integral part of British history rather than something that happened somewhere else. We need to stop 'turning a blind eye', knowing and not knowing, acknowledging the presence of empire but at the same time not confronting its meanings, especially its unsavoury ones. As Morrison comments in her discussion of the literary tradition of the United States, 'in matters of race silence and evasion have historically ruled literary discourse. Evasion has fostered another, substitute language in which the issues are encoded, foreclosing debate.' 'But', she continues, 'the situation is aggravated by the tremor that breaks into discourse on race'.

Let us now return to 'Cadbury World' and the tremors, the anxieties and the unsettled states of mind that disturb that narrative – not with the intention of pointing a finger of blame, for this is a history which implicates us all. Rather, the intention of the case study is to remind us of the ways in which empire is integral to daily life. Drinking a cup of tea, reading *Jane Eyre*, watching *Jewel in the Crown*, all of these remind us of the ways in which empire pervades British culture.

Memory pictures at an exhibition

'There's a lot more to Cadbury than making chocolate' proclaims a banner headline as we go into the first section of the exhibition. At the entrance visitors are given a hat made to the pattern of those worn by the women workers in the factory, to keep their hair well out of the way of chocolate production, and bearing the words 'Official Visitor to Cadbury World', together with a bar of chocolate to remind us of what the product is all about. We then enter a strange and tropical place with sounds of crashing thunderstorms and jungle noises where Mayan people in loincloths were, we are told, the first to grow and enjoy cocoa.

Cocoa merchants traded far and wide and a map of Central America shows us the trade routes which took the knowledge of the cocoa bean to the neighbouring Aztecs. 'The Aztecs were a very superstitious people who lived in an uncertain world constantly threatened by catastrophe' reads a caption. Catastrophe indeed struck them, in the form of Hernando Cortés, for the colonial encounter between Cortés and an Aztec woman Marina meant that he heard tales of the riches of Quetzalcoatl. This encounter 'signified the beginning of the downfall of the Aztec Empire and the discovery by the Europeans of the spicy drink known as chocolate'. Thus the 'native' woman, both the feeding mother of infancy and the sexual woman, gave the European man access to the New World and its delights. After the conquest of the Aztecs, Cortés was made Governor of Mexico and he returned to Spain in 1528, taking cocoa beans and chocolate-making equipment with him. At this point in the exhibition there is the opportunity to taste the spicy drink with its special recipe, a history consumed. Already colonialism has been clearly signified through its code word 'discovery', and the Spanish labelled as the conquerors who shattered the 'Old World' which they found in the 'New'.

The English are now introduced into the exhibition. The first advertisements for chocolate appeared in 1657. The drink probably came from France and became popular at the court of Charles II. It was described as 'an excellent West Indian drink' and one of the sources of supply in seventeenth-century London would have been the cocoa walks planted by the Spanish which Cromwell's troops found when they conquered Jamaica in 1655. Coffee and chocolate houses, associated with luxury and sexuality, were favoured places to partake of the new drinks and the façade of White's Chocolate House, on London's St James' Street, is reproduced in the exhibition.

Next door to White's Chocolate House is a reconstruction of Bull Street in Birmingham where the Cadbury story begins. At the end of the eighteenth and beginning of the nineteenth centuries, Birmingham was a fast-growing town needing new retail stores, and Richard Tapper Cadbury, a Quaker from Gloucestershire, established his linen draper's business in 1794 and opened his shop on Bull Street. In 1824 his second son John opened a tea and coffee shop next door while his older brother

Benjamin continued to run the original family business. In the exhibition these enterprises are presented in the time-honoured way, as stories of individual male entrepreneurship, successful against many odds. The family business becomes the male line and the efforts of Elizabeth Head Cadbury, the patriarch Richard's wife, or of the capital and labour which came through wives, mothers and daughters, the centrality of the networks of family and friends which women worked so hard to maintain, all of this disappears from the narrative. The Bull Street shop opens and takes us into two tableaux, representing in their very form the fixity of the historical past. One shows us John Cadbury in the cellar of his shop in the 1830s, experimenting with cocoa beans with his pestle and mortar, preparing small quantities of cocoa for sale. The second shows his two sons Richard and George, working in the shop in the 1860s with the new chocolate press from Holland which was to make the Cadbury fortune. Stacked up next to George are the crates of Ceylon tea which the business still depended on for its daily bread.

Tea, coffee and chocolate had been the commodities of the rich but were fast becoming part of the daily lives of a much larger percentage of the population. As the anthropologist Sidney Mintz comments, tea with sugar became part of the English 'character'. The tea at this time came mainly from China and Ceylon. One of John Cadbury's marketing ploys was to employ a Chinaman in native costume behind the counter to sell tea, for he was fully aware of the power of the 'exotic'. The tea then was from China and Ceylon, the coffee and cocoa mainly from the 'New World', from the West Indies and Central and South America (Figure 1). The Cadburys were well known as antislavery supporters, part of that distinctive abolitionist public, so it is likely that their sugar in the days after the end of slavery was bought from the newly emancipated British West Indies. The Cadbury women had been active in the campaign to boycott slave-grown sugar, a movement which started in Birmingham and was part of the adventurous strategies of the Birmingham Female Society for the Relief of British Negro Slaves.

From its beginnings, therefore, the Cadbury business was heavily dependent on the products of empire, whether that empire was British, French or Spanish. They were also seriously involved in efforts to change the relations of empire, to abolish slavery and encourage free labour.

Figure 1 'Europe supported by Africa and America'. An allegory of colonial relations by William Blake used as an illustration for John Stedman's *Narrative of a Five Years' Expedition, Against the Revolted Negroes of Surinam*, 1796. Stedman stated that all people are 'created by the same hand and after the same mould' but qualified this by pointing out that not all are equal in authority.

By mid-century cocoa was promising to become the key to any possible growth, and the new invention, soluble cocoa, is advertised in the window of the reconstructed shop on Bull Street. Its selling points were that it was cheap and quick to make, a drink for all families and all classes. This non-alcoholic drink, dear to the heart of temperance families, for the Cadburys also supported temperance, was from the start associated with the development of a healthy and happy domesticated family life.

The products came from empires. As if in peripheral recognition of that pre-eminence of empire, but without fully seeing it, the 'blind eye', a map of the 'Old World' hangs above John Cadbury's desk in the reconstructed shop, where he sat to do his accounts, with the British territories marked in red. The cultural critic Walter Benjamin's comment comes to mind:

> To articulate the past historically does not mean to recognize it 'the way it really was'. It means to seize hold of a memory as it flashes up at a moment of danger.

The map seems to act as the 'moment of danger', the red signifying the British Empire, yet so easily missed, just hanging there on the wall, the red territories reminding us of the connections between Cadbury and that imperial legacy. Yet the map does not even include the 'New World', the source of the cocoa that was to be crucial to the Cadbury venture, perhaps an example of the knowing and not knowing, seeing and not seeing, which is evoked by the 'blind eye'.

This sense of the knowing and not knowing haunts us as we move past an exhibit celebrating the creation of Cadbury's Dairy Milk, the product which sealed the success of the family business. A large map of the world entitled 'Cadbury – A Global Business' lights up the areas where cocoa and sugar are now produced, two of the three key elements of the chocolate. The areas of cocoa and sugar production and the siting of the factories contain an imperial story, but one that is implied, not told. Another example, perhaps, of the way in which empire haunts British national culture rather than occupying centre-stage. A story which at some level we all know, but never bring to full consciousness. Cocoa beans were taken in the nineteenth century to West Africa, taken from the 'New World' by European colonizers to that part of the 'Old World' from which they had in part peopled the 'New', to build new fortunes. Sugar, established in the British West Indies from the late seventeenth century with labour forcibly taken from Africa, had turned the islands of Jamaica, Barbados and the Antilles into monocultures, for the pleasure of Europeans who longed for sweetness in their drinks and puddings. Emancipation and the expansion of European sugar beet production disrupted those economies and left them as residues, condemned to the margins of the global capitalist economy. The sites of Cadbury factories in the rest of the world again tell us an imperial story. For the first factories were built in the white settler colonies. Gradually they spread to other parts of the Empire – Ghana, India, Kenya, Malaysia, Nigeria, Singapore and Ireland – for it was not only systems of administration and patterns of rule which the British spread across the Empire, it was also commodities and tastes.

Cadbury started selling abroad in the 1870s but their export market did not really get established until the 1880s when they sent representatives of the firm out into the wider world. Australia was the first outpost, part of the white settler world, 'Greater Britain'. Then came New Zealand and South Africa, India, Canada and the West Indies. The first depot was in Australia, the first factory a joint venture with Fry, another Quaker chocolate firm, in Montreal. The most successful export markets by the early twentieth century were Australasia, South Africa and India. National tastes in chocolate and confectionery vary greatly and Cadbury offered an especially British taste. Staying English in the Empire could be aided by eating Cadbury's chocolate.

Next to the map is a video display which tells that part of the imperial story which Cadbury likes to be told – the story of their support for the industrious peasant farmers of Ghana, once the Gold Coast, and the ways in which they have fostered West African development. Here the blind forces of fate in the shape of African 'backwardness', the laws of racial hierarchy and chaotic peasant production, were battled against by the Quaker firm working alongside the enlightened Colonial Office and the Gold Coast government prior to independence. By the early twentieth century, progressives had realized that it was essential to defend African land rights against white expropriation. The business was firmly aligned with progressive opinion in Britain and deeply critical of the forms of exploitation which some nations and businesses had engaged in. Cadbury took the most enlightened version of 'the civilizing mission' to West Africa, and are proud of that legacy. The continuously running video tells of the move to West Africa in the search for the best cocoa beans. Cadbury began to buy from the Gold Coast in 1907, fostering economic rationalization from the start by 'encouraging the farmers to adopt better methods' and then shipping the cocoa to Bournville where production of the chocolate took place.

The third wall in this triangular space focuses on milk – the English ingredient from the heart of the English countryside. This was the element which finally ensured the success of the chocolate bars – the most famous of which was Cadbury's Dairy Milk, made with fresh milk, first marketed in 1905. The Van Houten chocolate press, brought to England by Richard and George Cadbury, extracted the cocoa butter from the bean. That cocoa butter, mixed with sugar and added back into cocoa liquor, could be set into moulds and facilitated the production of chocolate bars. The addition of fresh cream milk made 'CDM' into the favourite it has remained, the most successful line ever. 'Why does Cadbury's Dairy Milk consistently come out Top in Consumer Tests?' asks a poster. The answer is that:

> Cadbury uses the world's best cocoa beans from West Africa.

To quote from the captions:

> The freshness and creaminess of milk straight from the farm is captured by combining the milk with the basic chocolate ingredient and with sugar, at a factory in the heart of the Herefordshire countryside . . .

English milk adds the vital white ingredient and the English countryside the vital healthful touch so that CDM can express the essence of Englishness in a chocolate bar, the 'basic' materials of which came from the erstwhile empire. Furthermore, the purity of the milk secures the inherent goodness of the product.

At this point the exhibition turns to Bournville and the life created there, followed by a brief factory tour. Our attention is now focused on the 'new world' of home and work – that world which reorganized and regulated class and gender relations, producing the worker in the factory and the consumer in the home. Richard and George Cadbury's success was predicated on their Quaker faith as applied to business. Hard work, dedication, lack of self-indulgence, abstinence from luxuries were the key, combined with the sense of a greater good, a better society. Business for them was a trust and workers had the first claim on an employer's benevolence, a paternal care. Cooperation should be the basis of business. The goods produced should be fairly priced and socially useful. Bournville, both new factory and village, was a response to this, part of the wider scheme associated with Joseph Chamberlain and the 'civic gospel' for the transformation of the city and its creation as a beacon for the nation.

In 1895 land next to the factory was bought for the site of Bournville Village, still there, with its well-spaced houses each with a good garden, elementary schools, day schools, churches and community centres. Bournville was the living symbol of the new world, a world of cooperation not conflict. Inequalities could be overcome, not by a socialist revolution and a change in patterns of ownership as many workers dreamed at this time, but by a sharing of the benefits of the enterprise, both at home and, to a lesser extent, in the Empire.

Two final sections of the exhibition deal with the marketing of the chocolate, from the earliest advertisements to the recent promos, and a special children's fantasy world which demonstrates the processes in the making of chocolate. Marketing was a preoccupation of the family firm from the beginning. Up to mid-nineteenth century, import duties on cocoa beans were high and kept chocolate as a luxury commodity. In 1853 Gladstone brought down the duties very significantly and this opened the way for cocoa becoming a genuinely popular taste.

While George Cadbury focused on the buying of the raw materials and the production process, his brother Richard put his mind to marketing and sales. Purity became the by-word of the firm, Cadbury's cocoa was 'absolutely pure, therefore best'. Newspaper and magazine advertisements focused on the absence of adulteration in Cadbury's cocoa, to the annoyance of other manufacturers. From the 1870s this healthy product was, furthermore, produced in a 'garden factory' and Cadbury concerns thus resonated with the concerns of the late nineteenth century as to the health and purity of the nation. Pure foodstuffs, with no extraneous 'foreign' materials, made by clean women and men in a model factory with homes, parks and welfare facilities attached could certainly claim a large section of the market. The rhyme, which was probably Richard Cadbury's own,

> Who wish for pure cocoa in all its quintessence
> Will certainly find it in Cadbury's Essence

was an inspired piece of advertising. Richard Cadbury also paid much attention to the boxes and hit on the notion of painted covers, using his drawings of his own children amongst others. The first image was of a charming little English girl in a muslin frock with a kitten on her lap and a red flower in her hair. The innocent and pretty little middle-class girl – presenting cups of the 'absolutely pure' soluble essence to the willing public, sitting with a kitten, skipping in the English countryside – was constantly re-presented.

Sidney Mintz in his book on sugar, *Sweetness and Power*, argues that

> Tobacco, sugar and tea were the first objects within capitalism that conveyed with their use the complex idea that one could *become* different by *consuming* differently.

Mintz is particularly interested in the new democratic identities associated with sugar. The good life was the sweet life and was available to all. Cocoa and chocolate were also constitutive of new identities, for the advertising linked these products to clean, warm and white worlds of happy families, where all children could be like the Cadbury children, healthy, pure and loved, sipping steaming hot cocoa or playing in the English countryside. In this process the 'gift' from the colonies played a crucial part. One advertisement, for example, shows a stereotyped black boy giving a stereotypically angelic white girl a cup of 'absolutely pure' Cadbury's cocoa, the gift of the colonies to Britain. Innocent black people, this image seems to suggest, love to give to white people. Darkness makes lightness, the black boy and the dark drink serve to highlight the whiteness of the little girl. In the process the colonies are domesticated too – black boys can occupy the same space as white girls, can be clean and bright. Furthermore, imperialism was domesticated, taken onto the breakfast table and consumed within the family. Cocoa is good for us, the good life was the pure life. Similarly a cocoa tin bears the words 'The gift of the colonies of Trinidad, Grenada and St Lucia.' Thus the relations of colonization are occluded, the labour of black people transmuted into gifts in return for the present of 'civilization'.

The wording of the poster next to the black boy and white girl reminds us again of the 'blind eye'.

> Cadbury's first association with the countries of the British Empire was its purchase of two small estates in Trinidad in 1897.
>
> Their important relationship with Ghana (or the Gold Coast as it was then called) began with a visit by William Addington Cadbury in 1909. The following year the first consignment of cocoa beans from Ghana arrived for processing.

The purchase of the two estates in Trinidad may have been the first purchase of land from the Empire – it was hardly the first association. The firm had been buying beans from the British colonies of Trinidad and Grenada for decades. Cocoa had been introduced into the Gold Coast in 1879 and by 1891 it was being exported. By 1929 it was producing 43% of the world output. In 1907 the firm sent an agricultural inspector to lay the basis for their buying from the Gold Coast and 'native farmers . . . were instructed how to plant, cultivate and prepare good marketable cocoa . . . they showed remarkable quickness in benefitting by the instruction given them'. Furthermore, cocoa growing was concentrated into the hands of men, for the marginalization of women growers was a part of colonial policy, re-creating the English sexual division of labour across the Empire.

The 1990s version of the colonial or postcolonial relation is signified for us in a video documenting the making of a recent advertisement for 'the lady loves Milk Tray'. Like most contemporary confectionery advertising, this focuses on the sexual, sensual and luxurious aspects of chocolate, back to the associations of the Coffee House – a shift partly associated with the shift in emphasis from cocoa to chocolate. Some 70% of confectionery sales are impulse buys, which makes advertising particularly salient because it is the symbolic meanings of it which are crucial to sales. This advertisement for Milk Tray took six months to make, with careful scripting, budgeting and casting. One week was spent in Jamaica filming the piece. The location had to be perfect and the Blue Mountains with their spectacular beauty evoke the James Bond connection. A 007-like character arrives against all odds with the chocolates. The lady who loves Milk Tray and her lover are both white, as were the director, producer and cameraman. One black man is visible with the film crew and another moves props. Thus Jamaica, the first site of colonial cocoa walks in 1655, is refigured as the exotic site for the selling of the product. The history of the beans, of the labour which produced them, of the unequal relations which characterized that production, of the racialized forms of representation which sold them, all are occluded.

Cadbury histories

Cadbury has been telling its own story for a long time and Cadbury World is only the latest version of that history. Indeed, the history has been an integral part of the marketing of Cadbury's products and the persistence of the Cadbury legend. The first publication was Richard Cadbury's *Cocoa: All About It*, published in 1892. This told, with illustrations, of the history and cultivation of the cocoa plant, the development of its use and the crucial part played by the Cadburys, for nearly one-third of the cocoa imported into England went to their business. The process of manufacture was described, the care that went into maintaining purity, 'no unsound berry or foreign material is passed into the roasting room', the division of labour in the factory with the young women hand-making the chocolates and finishing the boxes, the good employer/employee relations which were maintained particularly through the dominance of piece work. The story of Bournville featured too, with its first sixteen semi-detached houses built for the most prominent 'hands' in the factory. The structure of the 'Cadbury World' narrative was already there. The story was gradually developed and elaborated, produced in pictorial form, as a lecture with lantern slides, as educational materials, disseminated through the popular press.

A Cadbury publication in 1900, *From the Tropics to BRITISH HOMES via Bournville* told the picture in stories. The beans were grown and picked in Trinidad and Ceylon and brought to England.

> Ideal conditions of labour, careful supervision and absolute cleanliness
> are reflected in the wholesomeness, the delicious flavour, the purity of all

the Cadbury products, and in the fact that the name Cadbury is a
household word all over the world.

It was this cleanliness and purity, signified in the sweet young girl of the
advertisements, which made the products appropriate for the home. The
heat of the tropics and all that that could stand for – sexuality, excess,
blackness – was exorcised by the labour of an improved working class,
with their decent conditions of work, their uniforms, their proper homes
and gardens.

In 1909 a libel case rocked the firm's reputation. Cadbury went to
court asking for substantial damages from the *Standard*, which had
alleged that they were concealing their purchase of slave-grown cocoa.
Cadbury had been buying cocoa from São Tomé and Principe from the
1890s. From 1901 the management knew that conditions there
resembled slavery and began to investigate and to put pressure on the
Portuguese government to do something about it. Cooperating throughout with the Colonial Office and the antislavery organizations, they were
convinced until the end of 1907 that their best strategy was to continue
buying and thus exercise influence. In the libel case, the conservative
lawyer Sir Edward Carson accused the Cadburys of buying slave-grown
cocoa for eight years while selling their product on the basis of their good
treatment of their employees, their 'factory in a garden', the 'simple and
happy life' they promoted, in other words their 'absolutely pure' cocoa.
Cadbury's purity was bought at a price, Carson insisted. The trial
resulted in farthing damages for the firm. The case highlighted the
antagonisms between conservatives and progressives on 'the race question' and the difficult judgements which antislavery advocates had to
make.

'Once in a while', as the African-American theorist W. E. Du Bois
commented, 'a feud between the capitalists and manufacturers at home
throws sudden light on Africa'. Cadbury was hated by the Unionists
for attacking the Boer War and drawing particular attention to labour
conditions on the Rand. The Unionists were delighted to be able to
retaliate by attacking the Cadburys and suggesting that the firm supported 'black slavery'. The unwitting effects of this were that black
peasant proprietors became independent of English capital in São

Tomé, and black peasant farmers in Nigeria 'performed one of the industrial miracles of a century and became the centre of a world industry'.

The confident imperial voice which marks the publications of the early twentieth century disappeared as decolonization gathered pace. The postwar material on West Africa stressed the ways in which cocoa farming had enabled Africans to improve their standard of living and lead more settled lives. By the 1950s a schools publication, *Cocoa Farmers, Chocolate Workers*, reflected a new mood associated with the move to independence. Cadbury had civilized and modernized the Gold Coast. 'Improved' African small farmers could confidently occupy the same world, the same chain of production, distribution and consumption, as the chocolate workers of Bournville, with their access to the educational and welfare system of the twentieth century. Bournville is now celebrated as the crucible of modern industrial relations, cooperative, efficient, a partnership, the new forms for the late twentieth century, marked by equality, the British way forward. This remains the tone of 'Cadbury World'. The tales of empire are partially erased, yet the colonial relation is ever present. Interdependence between nations and peoples is the focus of the narrative. All the potential elements of another story are there, but the 'blind eye' is turned, as it is turned throughout British national culture.

Conclusion

Two vignettes bring this story to a close. In telling earlier versions of my narrative, personal stories from the audience have confirmed my sense of the ways in which Cadbury is embedded in popular memory and culture. In the late 1950s, on the eve of Jamaican independence, a young white English boy was dreaming of faraway places, of empire, of the doings of Winston Churchill and Robert Baden-Powell as recounted on the back pages of his favourite comic, *The Eagle*. Cadbury came to his school and a special filmshow was set up – it gave him his first taste of Jamaica as he saw the cane-cutters harvesting the sugar which went into the chocolate he loved to eat. White men organized the Empire he learned; black men laboured in the colonies. Twenty years later a young black British boy growing up in Leicester, a city with a high incidence of

South Asians and a significant 'race' problem, was racially taunted in the playground with, as he remembered,

> Nuts, whole hazlenuts,
> Cadburys take 'em and they cover them in chocolate

Cadbury, its products, its story, is part of the world of our raced imaginations, part of our tastes, part of our collective memories, part of our classed and gendered identities, part of how histories work in the present. At Cadbury World this raced history is uneasily occluded, provoking tremors in the discourse. A benign version of the past reigns, located around notions of modernity and progress, jarring with the raced memories of both the white English and the black British boys. If the members of Britain's current multiethnic population are to be full participants in the society of the millennium then blind eyes must be opened, for opening our eyes wider might enable us to see and understand differently.

FURTHER READING

Davidoff, L. and Hall, C., *Family Fortunes. Men and Women of the English Middle Class 1780–1850*, London: Routledge, 1987.

Mintz, S. W., *Sweetness and Power. The Place of Sugar in Modern History*, New York: Viking, 1985.

Morrison, T., *Beloved*, London: Picador, 1988.

Rowling, N., *Commodities. How the World was Taken to Market*, London: Free Association Books, 1987.

Steiner, J., 'Turning a Blind Eye: the cover-up for Oedipus', *International Review of Psycho-Analysis*, **12** (1985), part 2.

Williams, I. A., *The Firm of Cadbury 1831–1931*, London: Constable, 1931.

3   **Memory and the Making of Fiction**

A. S. BYATT

I read somewhere that writers, or perhaps all artists, are haunted or hunted by unusually vivid memories of their early lives – as are the old, whose useful daily memories are decaying. Certainly I remember being obsessed as a child by a kind of 'glittering' quality about certain experiences, usually without deep importance in what I thought of as the narrative of my life – experiences excessively bright, strongly outlined, *recognized* so to speak as important, even when they were met for the first time. (*Re-cognized* implies memory, the existence of a former cognition.) It is not too much to say that these experiences were as tormenting as they were delightful, until I, the person who underwent them, formed the project of being a writer – because only the act of writing gave them a glimmer of the importance they had in life, and thus gave them a place, a form and an order which made sense of them. And which they seemed to ask for. Proust, at the beginning of *À la Recherche du Temps Perdu*, speaks of certain experiences which *forced him to look at them* – 'a cloud, a triangle, a belltower, a flower, a pebble' – and gave him, as a child, a sense of *duty* towards them, a feeling that they were a symbolic language which he ought to decipher.

I shall come back to Proust, more than once, in the course of this essay. Now I want to give an example of a moment *recognized* in perhaps 1967 as part of a novel, which I have just, in 1995, written into one. In *Babel Tower*, which is the third part of a four-book series of novels, this memory is written down by Frederica, the central character, in a work she is making called 'Laminations', which contains cuttings from news, quotations from books, cut-up letters from lawyers, and odd disjunct records of her own. The memory – which is mine as well as hers – differs from Proust's involuntary recall of his childhood when he tasted a

47

madeleine, because it presented itself immediately to be *recognized* as a memory, at the time of being an experience. The mechanism by which it did this is, I think, as mysterious as that of the madeleine.

> A woman is sitting in Vidal Sassoon's salon, the Bond Street one. She is having her long hair, which she has always had, shorn into one of those smooth swinging cuts, like blades in their precise edges and points. Two young men are working together on the nape of her neck. Her feet are surrounded by shanks and coils and wisps and tendrils of what until recently was her body. It sifts, it is soft, it pricks between her collar and her skin. One man leans over her and holds the two points of her new hair down, dragged down, to her jaw. He hurts her. If she tries to look up, he gives her a little push down again, which hurts her. The other works above the vertebrae of her naked nape with his pointed shears. She can hear the sound of hair on blade: a silky rasping. He nicks her skin with his points. He hurts her. She is almost sure that the small hurts are deliberately inflicted. Over her head the two talk. 'Look at that one then, look at her strut, she thinks she's the *bee's knees*, the *cat's whiskers*, and she's a walking *disaster*, look at that clump he's done at the back, like a great *bubo* all bulging and she can't see it, she can't see it jiggle and wiggle as she walks, she thinks she looks *delicious*, he told her so, he held the mirror at the right angle, so she couldn't see what a godawful *mess* he had made, cutting higher and higher trying to make it better and now there's nothing left to cut, only a gob at the back of her *lumpy head*.' They laugh. The woman under their hands tries to look up and is jerked down. She thinks, I will always remember this, but doesn't know why; there are many humiliations, many disasters, why will she always remember this one? They let her head up. She sees her face through tears. The line is like a knife along her jaw. They tell her she looks lovely. All the women in the room have the same cut and all look lovely in the same way, except those who don't. When she moves her head, the curtain of her hair swings and re-forms into its perfect edge. Her neck is naked. She gives the two a tip, though she would like not to. Her hair looks good. Does she?
>
> (Byatt, 1996, pp. 388–9)

This account differs from the actual mnemonic in various ways. What is missing is the geometry of the light of the new 'open-plan' salon, the multiplied mirrors which haunt a disproportionate number of my important memories. What is also missing is a primitive fear of lack of privacy which is one of my most persistent emotions. Although the young men did mock the parting customer, the images for her shearing are my own, verbal reconstructions of an immediate visual impression. I have tied (in 1995) my memory tightly to the meditated cultural themes of the novel's portrait of the sixties – the note of gay and gleeful cruelty in all the new freedoms, the absurd uniformity of all our attempts to look new and different, the sense that sex was still dangerous for women. I have also tied it, in the long metaphorical structure of the novel, to the theme of pairs of men, starting with Kafka's two 'assistants' in *The Castle* who were connected by a student in my extramural class with twins and with testicles, and ending with newly invented identical twin brothers (newly in the 1990s) who are in their turn connected with Nietzsche's Apollo and Dionysos. And partly beginning from there, and from a memory of a sculpture student called Stone who fell under an underground train whilst stoned, I have found the game of 'scissors, paper, stone' patterning my text, and having found it, have elaborated it. I knew moreover, both immediately and in the writing, that the 'shears' are those of Milton's blind fury who slits the thin-spun life – the fear of death is and was in both the experience and the text. But I only noticed on typing out my passage how much it is to do with the fear of execution, of beheading – Frederica played Elizabeth I in *The Virgin in the Garden*, and I was meditating at the time of the haircut on virginity as a defence, a source of power, of not being cut, or harmed, or killed.

Personal memories

The great modernist writers – Proust, Joyce, Mann, Woolf – weave significant personal memories, recognitions, epiphanies, into a texture of language and thought. The philosopher Richard Rorty sees Proust's projected novel as an exemplary form of a modern, provisional, *ironist* culture, which has given up the search for eternal and immutable laws or theories of human nature. He compares and contrasts Proust with Nietzsche. Both men, he says, believe in self-creation by self-description, in becoming one's 'true' self by redescribing one's world.

> For Proust and Nietzsche there is nothing more powerful or important than self-redescription. They are not trying to surmount time or chance but to use them. They are quite aware that what counts as resolution, perfection and autonomy will always be a function of when one happens to die or go mad. But this relativity does not entail futility. For there is no

A. S. Byatt

big secret which the ironist hopes to discover, and which he might die or
decay before discovering. There are only little mortal things to be
rearranged by being redescribed.

(Rorty, 1989, p. 99)

Rorty sees Proust's novel as wiser than the systems of the ironist theorists in whom he is interested – Hegel, Heidegger and Kant. 'Proust's
novel', he says,

is a network of small interanimating contingencies. The narrator might
never have encountered another madeleine. The newly impoverished
Prince de Guermantes did not have to marry Madame Verdurin: he
might have found some other heiress. Such contingencies make sense
only in retrospect – and they make a different sense every time
redescription occurs. But in the narratives of ironist theory, Plato *must*
give way to St Paul, and Christianity to Enlightenment. A Kant *must* be
followed by a Hegel, and a Hegel by a Marx. That is why ironist theory is
so treacherous, so liable to self-deception.

(*Ibid.*, p. 105)

Nietzsche, Rorty says, was tempted to identify himself with abstractions
like Will to Power, Becoming and Power.

Proust had no such temptation. At the end of his life he saw himself as
looking back along a temporal axis, watching colours, sounds, things,
and people fall back into place from the perspective of his own most
recent description of them. He did not see himself as looking down upon
the sequence of temporal events from above, as having ascended from a
perspectival to a non-perspectival mode of description.

(*Ibid.*)

Proust makes a pattern from the little things – Gilberte among the haw-
thorns, the colour of the windows in the Guermantes chapel, the two
walks, the shifting spires. Rorty continues,

He knows this pattern would have been different had he died earlier or
later, for there would have been fewer or more little things that would
have had to be fitted into it. But that does not matter. Beauty, depending
as it does, on giving shape to a multiplicity, is notoriously transitory,
because it is likely to be destroyed when new elements are added to that
multiplicity. Beauty requires a frame, and death will provide that frame.

(*Ibid.*)

I feel very happy with the idea that the novel is a profoundly agnostic and provisional form, connected to local truths and making sense, in the shifting structure of a whole life, of those disproportionately glittering perceptions and events I began with. But *self*-description or re-description is not an adequate account of the way memory works in Proust's novel – and the personal memories of writers are haunted by, and connected to, the memories of the dead, both the immediately dead and remembered, and the long-dead whose memories constructed the culture we live in and change in our turn. George Eliot's fiction springs very powerfully out of her own personal memories – we know as readers of *The Mill on the Floss* that the child Maggie Tulliver, with her passion-ate attachment to her elder brother and the landscape of her upbring-ing, grew from the child Mary-Ann Evans. What is more interesting is that Eliot's fiction appears to progress chronologically through the remembered life, not only of the writer, but of her father and her family. *Scenes of Clerical Life* combines her earliest memories of accompanying her father to Arbury Hall with her father's memories and stories of events before her own lifetime; in *Adam Bede*, she bases Adam on her father, and then universalizes him by calling him Adam, and making him, like Christ, the new Adam, the son of a carpenter. She wrote in Chapter 3 of *Daniel Deronda* of the importance of memory in the con-struction of a life:

> Pity that Offendene was not the home of Miss Harleth's childhood, or endeared to her by family memories! A human life, I think, should be well rooted in some spot of a native land, where it may get the love of tender kinship for the face of earth, for the labours men go forth to, for the sounds and accents that haunt it, for whatever will give that early home a familiar unmistakable difference amidst the future widening of knowledge: a spot where the definiteness of early memories may be inwrought with affection, and kindly acquaintance with all neighbours, even to the dogs and donkeys, may spread not by sentimental effort and reflection, but as a sweet habit of the blood.

This passage looks back to Wordsworth, and forward to Proust. Its sense of the importance of the local and the particular is connected to Wordsworth's haunting 'spots of time' by the word spot: Eliot's memory as 'a habit of the blood' is related to Wordsworth's 'felt in the blood, and felt along the heart' (*Tintern Abbey*, line 29). It prefigures Proust's involuntary memory, which is not an effort of will, but an experience of immediate sensation. Proust was a devoted reader of Eliot. I should like to draw attention also to the passage which speaks of 'the labours men go forth to, for the sounds and accents that haunt it', which contains a biblical echo from Ecclesiastes, a generalization about the repetitions of generations of human lives, and the verb 'haunt', which I discovered was haunting all my thoughts when I first began to meditate on memory and fiction. For particular, local personal memories seem immediately to evoke, in a kind of de Quinceyan involute, the idea of the dead and their memories, our own dead and the long-dead, whose memories haunt poetry.

It is interesting, if we look at Proust's first account of the discovery of involuntary memory, to see how he places metaphors of death and haunting (which are also facts as well as metaphors) just *before* the magic tasting. He says that the memory of intelligence could only have reconstructed a kind of schema, or décor, from his childhood, and then goes on

> Tout çela était en réalité mort pour moi.
> Mort à jamais? C'était possible.
> Il y a beaucoup de hasard en tout ceci, et un second hasard, celui de notre mort, souvent ne nous permet pas d'attendre trop longtemps les faveurs du premier.
>
> Je trouve très raisonnable la croyance celtique que les âmes de ceux que nous avons perdus sont captives dans quelque être inférieur, dans une bête, un végétal, une chose inanimée, perdues en effet pour nous jusqu'au jour, qui pour beaucoup ne vient jamais, où nous nous trouvons passer près de l'arbre, entrer en possession de l'objet, qui est leur prison. Alors elles tressaillent, nous appellent, et sitôt que nous les avons reconnus, l'enchantement est brisé. Delivrées par nous, elles ont vaincu la mort et reviennent vivre avec nous.
>
> Il en est ainsi de notre passé. C'est peine perdue que nous cherchions à l'évoquer, tous les efforts de notre intelligence sont inutiles. Il est caché hors de son domaine et de sa portée, en quelque objet matériel (en la sensation que nous donnerait cet objet matériel) que nous ne soupçonnons pas. Cet objet, il dépend du hasard que nous le rencontrions avant de mourir, ou que nous ne le rencontrions pas.
>
> (Proust, *À la Recherche du Temps Perdu*, vol. 1, p. 44)

Translation

To me it was in reality all dead.

Permanently dead? Very possibly.

There is a large element of hazard in these matters, and a second hazard, that of our own death, often prevents us from awaiting for any length of time the favours of the first.

I feel that there is much to be said for the Celtic belief that the souls of those whom we have lost are held captive in some inferior being, in an animal, in a plant, in some inanimate object, and so effectively lost to us until the day (which to many never comes) when we happen to pass by the tree or to obtain possession of the object which forms their prison. Then they start and tremble, they call us by our name, and as soon as we have recognised their voice the spell is broken. We have delivered them: they have overcome death and return to share our life.

And so it is with our own past. It is a labour in vain to attempt to recapture it: all the efforts of our intellect must prove futile. The past is hidden somewhere outside the realm, beyond the reach of intellect, in some material object (in the sensation which that material object will give us) which we do not suspect. And as for that object, it depends on chance whether we come upon it or not before we ourselves must die.

(Proust, *Swann's Way*, p. 45)

Our first worship, Sigmund Freud says, was ancestor-worship. Our immediate ancestors *are*, in some sort, our memories – they disappear, as presences, as bodies, but they persist as icons, as hauntings, in our minds. They take with them their memories of ourselves, so that part of us dies with them, as part of them persists in us, both in the genes and in the workings of the brain, both ways. I at least, and I think this is common, became interested in my more distant ancestors at the time of the death of my parents. It was then that I began writing ghost stories, taking ghost stories seriously, and also using a metaphor of ghosts as the life-in-death of poets in poems, of poems in my memory. Poems may live in the mind as icons – I think of *Paradise Lost* and 'see' a complex map of colours and gardens and rivers and flying dark and light creatures – or in language learned 'by heart', so that when I embark on reciting a Shakespeare sonnet I am confident of being able to feel my way to the end, word by word. It is at the point when one begins to feel one's own mortality – as Proust was feeling his when he discovered his vocation –

that one becomes able to see not only one's finite life in the world, but one's own peculiar body of connected knowledge (knowledge is memory and mnemonics) as part of the order (provisional) of a work of art. What cannot be written will die with us. And at this point the connection to the great works of the past becomes urgent and personal, a bright discovery, a re-membering of countless other dead or dying memories.

When my own father died I realized that what had died with him was the store of his memories – his memories of his own father and mother, whom I hardly knew, my secondary, or inherited memories. He seemed to realize that this mattered; he talked about his parents on his death-bed as he had never done earlier. I wrote a story about his dying, *Sugar*, in which I tried to be truthful, to put together, to give form and beauty to, the accidents of his death, as I remembered it, and of his and my life, as they presented themselves in 'glittering' significant moments. It is inter-esting that into this purely personal, carefully factual account, a kind of mythic landscape did extend. It is a fact, though also an unnerving coinci-dence, that at the time when my father was collapsing on a Rhine cruise, I was watching the *Götterdammerung* at Covent Garden. This led me, in my story, to a true memory, my compulsive reading and re-reading, as a child, of Norse myths, my interest in the underworld and the Last Battle, Ragnarök, and the paradisal world that came to be after the death of the gods. That was juxtaposed with another fact, also a coincidence, that I was at the time working on the paintings of Van Gogh, *The Sower* and *The Reaper*, in which the painter mixed precise observation of corn and clods with ancient images of the labours men go forth to, and the reaping Death in a furnace of light. The historical accident of my father's dying in Amsterdam made it possible for me to study these icons during the days of his vanishing. And that again seemed somehow to connect with my father's dying memories of his own childhood as paradisal, or Arcadian, a time which smelled of horses and fresh grass and there were what he referred to as 'real apples and plums'. I begin to wonder if the forms of particular memories – at least in works of art – tend to call up a penumbra of family memories, ancestral memories, and if Paradise and the Underworld are images of this recollection, recalling of ghosts, re-membering of the dead and the vanished.

Oral poetry

I have been interested for a long time in the formal structures of oral poetry, about which I *know* very little, the kennings and repeated rhythms of Anglo-Saxon verse, for instance, which Michael Alexander says provide a 'special sort of improvisation-aid, a poetic grammar or rhetoric which acted as scaffolding for their song, infinitely flexible, extensible, adaptable to the job in hand.' Whilst I was informing myself about these structures, I found I was also making discoveries about the subject matter of oral poetry which seemed to relate directly to my sense of Proust's subject matter, and George Eliot's and my own. Oral poetry, several scholars point out, takes place in the present, a perpetual present – Virgil's *written* poetry recalls the written past of Homer, as Dante recalls Virgil, and Milton recalls all three. But Homer's world, supposing his poem was an oral poem, is *now*, and happens again every time the poem is sung – a poem which has not been learned, but is made new each time it happens, within the frame of rhythms and metaphors and names which are its skeleton. Jeff Opland, in an interesting book on Anglo-Saxon oral poetry, includes an excursus on Xhosa and Zulu oral poetry. The Xhosa *izibongo* he tells us, includes poems which most people know, about their clans, poems transmitted in fixed form from generation to generation. 'Heads of households also know the poems treating their departed ancestors, which are recited on ritual or cer-emonial occasions, in order to initiate communication between the living and the dead' (Opland, pp. 19–20). Most men seem to know the personal poems about their father and their paternal grandfather, Opland says, but not much further back, though they know the historical poems of the clan. Boys compose their own autobiographical poems as they grow up – from the accidents of their individual lives – as Proust composed his book, which was the redescription and purpose of his life. Boys learn their fathers' autobiographical poems, and use them as a ritual method of communication with their fathers after their death. Here are clan memory, immediate ancestral memory, and personal memory, for-malized and recited, in a way I think can be related to what Eliot is doing in the movement of her fictions.

Ronald MacDonald opens his book on epic underworlds in Virgil,

Dante and Milton with Macaulay's observation that Milton's poetry 'acts like an incantation':

> Its merit lies less in its obvious meaning than in its occult power. There would seem, at first sight, to be no more in his words than in other words. But they are words of enchantment. No sooner are they pronounced than the past is present and the distant near. New forms of beauty start at once into existence, and all the burial places of memory give up their dead.
>
> (MacDonald, 1987, p. 7)

Milton's language produces the same effect as Proust's magic madeleine – the past is present, and the burial-places of memory give up their dead. In the Odyssey, the hero visits the land of the dead, meets the shades of his mother and of the seer, Tiresias, and hears a prophecy of his future. In the *Aeneid*, the hero also visits the Underworld, accompanied by the sybil, or prophetess, and meets the shades of his dead companions, of Dido who turns away from him into the grove, and of his father, who tells him also a narrative of the future, of the Roman state he will found. In the *Divine Comedy* Dante is guided through the Inferno by the shade of Virgil, and meets both classical, Christian and mythical ghosts – including Ulysses, and Judas – as well as personal friends and enemies of his own, who have the kind of local life and idiosyncracy that Proust's society figures, and that his friends and enemies are to have. *Paradise Lost* is a later story which claims that it is earlier than the others, and takes place before there are any human dead to have any memories. The memories the reader shares are those of Satan, excluded from Heaven, an eternal inhabitant of the underworld prepared for the sinners, peering in at Paradise, whose beauty renders him briefly 'stupidly good', *imagining* the life in the walled garden from which he has always been excluded. Dante's Gate of Hell proclaims itself older than all created things which are not eternal – it is older than Virgil's mythic Avernus; however, the Christian texts which claim priority work by *remembering* their predecessors and ancestors. Consider, for instance, Milton's image for the fallen angels

> who lay entranced
> Thick as autumnal leaves that strew the brooks

> In Vallombrosa, where the Etrurian shades
> High overarched embower –
>
> (*Paradise Lost*, I, lines. 302–5)

These lines remember Virgil's description of the ghosts crowding the banks of the Styx stretching out their arms in vain longing for the other shore 'thick as the leaves of the forest that at autumn's first dropping fall':

> quam multa in silvis autumni frigore primo
> lapsa cadunt folia . . .
> stabant orantes primi transmittere cursum
> tendebantque manus ripae ulterioris amore
>
> (*Aeneid*, VI, lines 309–14)

Wordsworth's *Prelude*, or 'poem on my own life', is haunted by the words and ambitions of all these predecessors, but finds its significance in the contemplation and interrogation of his *particular, personal* memories – substituting spots of time, and glittering visions of real mountains – for angelic visitors, and the visionary dreariness of a woman struggling against a wind 'with her garments vexed and torn' for ghosts. Or invoking books. Book V of *The Prelude* is called 'Books' and contains both meditations on fairy tales and myths in the construction of a mind, and several summonings of the dead – the drowned man who rises upright from Windermere, Wordsworth's dead mother, and the dream of the fleeing Arabian knight with his symbolic stone and shell, running from the deluge – the fleet waters of a drowning world – with a stone that represents Euclid's Elements, and a shell that utters,

> in an unknown tongue
> Which yet I understood, articulate sounds
> A loud prophetic blast of harmony;
> An Ode, in passion uttered, which foretold
> Destruction to the children of the earth
> By deluge, now at hand.
>
> (*Prelude*, V, lines 91–7)

Wordsworth's sense of his self-construction is behind George Eliot's, and hers plays a part in Proust's, as I have said. It is interesting to find how *naturally* Paradise and the Land of the Dead occur at the point where Proust's narrator, in *Le Temps Retrouvé*, makes his discovery of the mechanism of involuntary memory which reveals to him the nature of the book to come which itself gives meaning to his life. The narrator is undergoing his own experience of 'visionary dreariness'; he has seen a beautiful row of trees but, like the Coleridge of the *Dejection Ode*, he 'sees not feels' how beautiful they are. He is going to a matinée at the house of the Prince de Guermantes, and his carriage turns into streets which go to the Champs Elysées.

> They were very badly paved at this time, but the moment I found myself in them I was, none the less, detached from my thoughts by the sensation of extraordinary physical comfort which one has when suddenly a car in which one is travelling rolls more easily, more softly, without noise, because the gates of a park have been opened and one is gliding over alleys covered with fine sand or dead leaves . . .
>
> (Proust, *op. cit.*, VI, p. 206)

The name, Champs Elysées, Elysian Fields, is a fact of Parisian history and geography, and Proust's account is precise and factual as to time and place. But the detail of the carriage on soft sand, and the dead leaves, recall the golden sand of the arena where the fortunate souls disport themselves in Virgil's Elysian fields, and the dead leaves, not insisted on, are surely the leaves of Homer, Dante, Virgil and Milton's Vallombrosa. The narrator recalls his past, as he travels, 'gliding, sad and sweet'. He meets M. de Charlus, who has had a stroke, but whose memory is intact, and who speaks to him rather like Wordsworth's shell.

> And the traces of his recent attack caused one to hear at the back of his words a noise like that of pebbles dragged by the sea. Continuing to speak to me about the past, no doubt to prove to me that he had not lost his memory, he evoked it now – in a funereal fashion but without sadness – by reciting an endless list of all the people belonging to his family or his world who were no longer alive, less it seemed with any emotion of grief that they were dead than with satisfaction at having survived them . . . 'Hannibal de Bréauté, dead! Antoine de Mouchy, dead! Charles Swann, dead! Adalbert de Montmorency, dead! Boson de Talleyrand, dead! Sosthène de Doudeauville, dead!' And every time he uttered it, the word 'dead' seemed to fall upon his departed friends like a

> spadeful of earth each heavier than the last, thrown by a grave-digger
> grimly determined to immure them yet more closely within the tomb.
>
> *(Ibid.*, pp. 211–12)

The Baron's Elysian lists of the dead are surely related to Homer's lists of dead heroes, which Virgil repeated word for word, giving them extended life-in-death, though the Baron is determined that the dead shall not be resurrected. The narrator, however, continues on his way, and in the courtyard of the Guermantes house has an experience that repeats the sudden bliss of the experience with the madeleine, when he trips on an uneven paving-stone.

> The emotion was the same; the difference, purely material, lay in the
> images evoked: a profound azure intoxicated my eyes, impressions of
> coolness, of dazzling light . . .
>
> *(Ibid.*, p. 217)

He tries to seize the 'dazzling and indistinct vision' as it passes, and realizes it is Venice, which he has tried earlier to preserve with deliberately memorized 'snapshots' (the French for snapshot is *cliché*) but which is now recalled as a complete sensuous atmosphere by the involuntary recall triggered by the uneven stones. He finds in the experiences of Combray-madeleine and Venice-stones 'a joy which was like certainty, and which sufficed without any other proof, to make death a matter of indifference to me'. A few moments later a servant knocking a spoon against a plate, and then the wiping of his mouth with a too stiff napkin, recall both the row of trees which had previously seemed lifeless (through the connection with the tapping of a hammer against the wheel of the railway carriage from which he had seen the trees) and the sea at Balbec, seen from his hotel as he wiped his mouth with an analogous cloth. These experiences make the past vividly alive in the present. The narrator thinks hard about why and how these memory-experiences are so full of happiness and significance and comes to the conclusion that it is because all our past experiences are *separate*,

> immured as within a thousand sealed vessels, each one of them filled with
> things of a colour, a scent, a temperature, that are absolutely different
> one from another, vessels moreover, which being disposed over the

whole range of our years, during which we have never ceased to change if only in our dreams and our thoughts, are situated at the most various moral altitudes and give us the sensation of extraordinarily diverse atmospheres.

(Ibid., p. 221)

And, Proust seems to be saying, if by hazard we connect – through a madeleine or a stone – to one of these *separate* atmospheres and places

for this very reason it causes us to breathe a new air, an air which is new precisely because we have breathed it in the past, that purer air which poets have vainly tried to situate in paradise and which could induce so profound a sensation of renewal only if it had been breathed before, since the true paradises are the paradises we have lost.

(Ibid., pp. 221–2)

He speaks in an exalted yet precise language of the effect of these retrieved memories on his consciousness. What he repeatedly calls the *resurrections* of buried time move him, he says, by 'the miracle of an analogy' to a moment, a timeless moment outside time, where death is not fearful, and the work of art can take shape. He uses the word *'chainon'*, connecting link, in a bridge which the unconscious memory finds it can throw between the separate summits of the landscape of the past: I want to come back to the microcosm of 'linked analogies' (to use Melville's term) or the 'hooks and eyes of memory' (to use Coleridge's) shortly. But I should like to observe, before leaving the passage we have been thinking about, that the Narrator's discovery of his Paradise Lost is followed by the Prince de Guermantes's matinée which is a kind of grimly tragi-comic vision of the Underworld, where the *haut monde* are undergoing metamorphoses before sinking into death. They become their ancestors as their faces become those of their parents or relatives; their places and genealogies forget them. Proust himself evokes the Homeric Nekuia, or meeting with the ghosts. Odette appears to have kept her youth –

And yet, just as her eyes appeared to be looking at me from a distant shore, her voice was sad, almost suppliant, like the voice of the shades in the *Odyssey*.

(Ibid., p. 323)

One friend is recognizable only by his laugh:

> I should have liked to recognise my friend, but like Ulysses in the
> *Odyssey*, when he rushed forward to embrace his dead mother, like the
> spiritualist who tries in vain to elicit from a ghost an answer which will
> reveal its identity, like the visitor at an exhibition of electricity who cannot
> believe that the voice which the gramophone restores unaltered to life is
> not a voice spontaneously emitted by a human being, I was obliged to
> give up the attempt.
>
> <div align="right">(<i>Ibid.</i>, pp. 314–15)</div>

Babel Tower

I began with a quotation from *Babel Tower*, the third in a series of novels
which began with *The Virgin in the Garden*. I think I began writing a
long novel because, having read Proust, I saw it as a solution to the problem of being haunted by 'glittering' memories which appeared to solicit
a place, a hearing. When I began the first novel I must have been in my
middle twenties, very conscious of being, as I then thought, old, and
having for the first time behind me a piece of public history *lived
through* – the Second World War, the 1950s, the time of the Festival of
Britain and the Coronation of Elizabeth II, of the coming of television,
which I already knew would change all our apprehensions of our experiences and our world, would form and reform all our memories and
methods of memorizing in then unimaginable ways. Proust's novel was
art coextensive with his life.

One of the experiences which formed my schooldays was the first performances of the York Mystery Plays, in which I failed to get a part, but
which involved almost everyone in York at the time in excited cultural
activity. I decided to write my novel about a performance of a play, in
verse, about Elizabeth I, which would involve a whole community, and I
thought I was interested in plays and pageants because of their cultural
inclusiveness. But looking back, I am sure it was partly because the Mystery Plays, enacted in the ruins of St Mary's Abbey, had a setting which
spoke to some primeval schema for human narrative with which I was
already working. (I had the great good fortune to have *Aeneid* VI and
Paradise Lost IX and X among my set books for A Level, as well as *King
Lear* and Mann's Dionysiac/Apollonian *Tonio Kroger*.) The Mystery

A. S. Byatt

Plays had heaven in the ruined Gothic arches, and a Middle Earth built between Heaven and the Mount of Hell where stood, as Donne said they were said to, Adam's Tree and Christ's Cross in one spot of time and place. I was obsessed by Paradise Gardens, by the topography of this religious theatre. I failed to write a thesis on temptations in Gardens from Spenser's 'Bower of Bliss' to 'Paradise Regained'. But I did study the iconography of Elizabeth I, especially in the works of that great scholar Frances Yates, who found that the Virgin Queen had managed to substitute herself, or be elegantly substituted by her poets, for both the Queen of Heaven (the Virgin cast out of English cathedrals in the days of Edward VI's puritan iconoclasm) and for Virgo-Astraea (a paradoxical goddess of justice and fecundity). Elizabeth had managed to be born under the harvest-time sign of Virgo, who can be related to Artemis and Astarte, the turret-crowned Cybele, the goddess of the groves, queen and huntress, chaste and fair. I discovered the Darnley portrait, that white-faced icon (Figure 1). I discovered Spenser's Dame Nature, who is both male and female and therefore self-sufficient and creative, and made myself a figure in a garden who represented female power and creativity – and separateness, an essential part of Elizabeth's survival. What interests me is that I almost immediately connected my white-faced human divinity to Keats's undying Moneta in *The Fall of Hyperion* – a figure, related to Aeneas's Sibyl and Dante's Virgil, who initiates the poet into awful secrets – and distinguished between the poet and the mere dreamer. Moneta is life-in-death. She resembles Proust's dying gods or aristocrats who have become their own effigies. In my novel I made the dying Elizabeth (who refused to go to bed) arrange the flutes of her gown into her own effigy. Keats's unfinished dream-vision *The Fall of Hyperion* is a later version of the also unfinished epic, *Hyperion*, which tells of the fall and mortality of the Titans, and in which Moneta's predecessor is Mnemosyne, the goddess of Memory.

Then saw I a wan face,
Not pined by human sorrows, but bright-blanched
By an immortal sickness which kills not;
It works a constant change, which happy death
Can put no end to; deathwards progressing

62

Figure 1 The famous Darnley portrait of Elizabeth I, by an unknown artist, illustrates how she became an iconic figurehead open to multiple interpretations.

To no death was that visage; it had passed
The lily and the snow; and beyond these
I must not think now, though I saw that face

(The Fall of Hyperion, lines 256–64)*

I found I was patterning my novel on a series of metaphors of white (death, stone, effigy), red (blood, violent death, also life, also loss of virginity) and green (the garden, the resurrection of the grass every spring). These fundamental markings, analogies, recurrent spots of colour, were mnemonics in the texture of my text, like patterns in embroideries or knitting, holding together the very personal and the largely cultural, the factual and the mythic. They were, as Proust says about the form of his novel, there *to be discovered* like the 'laws' of nature, patternings ambiguously 'out there' and present only in my own head, my own brain, my own life.

Memory structures within texts

I began writing the novel at a time when writers were very afraid that language was only a self-referring system, without any real link to the outside world. This leads to the idea of memory *within* a work of art, constructing a text. I remember being at a seminar of Frank Kermode's, in the early days of literary theory in England, when he remarked that it was impossible to hold the whole of a long novel in one's head, so that any account a critic gives of a novel must consist of certain stressed themes, or highlighted relationships, or spotlighted metaphors. I remember thinking that this both was and was not true – if you are writing a long novel there is a sense in which you do, precisely, *remember* all of it, including in some sense the part which is not written, which feels like a projected memory-image or mnemonic, as the whole of *À la Recherche* must have felt to Proust at the time of its conception.

This is a difficult thing to write about, and yet the fascination of the interlocking systems of mnemonics that constitute a novel in progress is one of the reasons I agreed to write this essay. I have never been quite able to understand the paradigm (now being rejected, as far as I understand, by cognitive theorists) of the 'picture in the head' surveyed by a homunculus, representing either consciousness or memory. My own most basic metaphors for this process have no homunculus. One is of feathers – being preened, until the various threads, with their tiny hooks and eyes, have been aligned and the surface is united and glossy and gleaming. The other is of a fishing net, with links of various sizes, in which icons are caught in the mesh and drawn up into consciousness – they come up through the dark, gleaming like ghosts or fish or sparks, and are held together by the links. Both of these descriptions are metaphors for something I do *feel* going on in my blood and brain and nervous system when I sit down to try to work. Memory within a text is always of two sorts, even when there is only one paragraph of it. There is memory of its own contents and coherence, and there is memory of the outside events – the madeleine, and beyond the madeleine the walled garden of Combray, and beyond the walled garden of Combray the walled garden of Paradise Lost – which are or will be part of this particular text. There are – before thought structures – rhythmic memories, and iconic visual memories, which must be allowed to form and be preened, so to speak.

Rhythmic memories are to do with the blood, and with language. I myself learn easily 'by heart', along the blood, and perhaps my most primitive and profound delight is to hear the rhythms of grammar and cadences of plosives and gutturals and tones working with and against the breath and the heartbeat. This is why I mentioned the mnemonics of the oral poets – I was delighted to read in one of those books of the dactyl being a long and two short joints, measured on a dactyl or a finger, so that Homer's rosy-fingered dawn is rhododactylos, a finger-mnemonic for a beat. We all experience the torment and delight of searching for something we *know* we know and have forgotten. With verses I usually find the significant shiny words by humming the rhythm, and finding the grammatical pointers, and with prose there is something of the same thing. I find that writing – more particularly verse, but prose too – is not unlike remembering: a rhythm begins to sing, a grammatical structure begins to hold, and the 'right' precise words take their places in the structure, like fish in a net, of which the knots are the small words and the rhythm. Rhythmic memory is knitting, and it extends over long passages marked with markers I can hardly describe – they are in a way *buried colours, the ghosts of colours* – twenty pages back there was a dark purple bit that needs a balance here, I think, a flash of yellow, a lighter purple, a dark gash, a run of pale marks.

This may sound like iconic memory, rather than bodily or blood memory. They overlap – the experience of looking at and memorizing a painting is rhythmic, a memory of scanning and saccades of the eyes, but the image of constructing a geometric icon is different. I used to have an almost eidetic memory – I could semi-visualize whole pages of text and 'read' them off, and I still know where, geometrically, something I am searching for, read or written, is on the page. But also one builds large static structures of mnemonics to put things into, to remember their relations. My own structures are rather like abstract paintings – a rising series of increasingly acute triangles in complementary colours may represent one text in construction, a series of concentric spirals, or even a double helix, another. My study of Frances Yates for Elizabeth brought me up by accident against the Elizabethan memory theatre, which derived from mediaeval systems for memorizing rhetorical arguments and facts. I think all theatres are memory theatres: one of the reasons one does not like to see films of novels one loves is the profound interference with the mnemonics of the novel caused by the image of the film, whereas no performance of King Lear, good or bad, disturbs our memory theatre of it in the same way. The Renaissance rhetoricians remembered arguments by placing them in visual arenas – sometimes in real places where they had stood and deliberately memorized. What interests me particularly about Frances Yates's arcane discussions – which range from astrology and cosmology to memorizing a ram with large testicles to remind lawyers of an opponent and his witnesses (testes) – is that both heaven, Paradise and Hell were pictured as memory systems, as stages, as internalized *frames* for remembering our lives (Figure 2).

My profound memory of the set of the York Mystery Plays feels in this context like a race memory, a kind of *grammar* of constructing narratives of our lives and deaths, endlessly transmutable. I connect it to the biologist E. O. Wilson's belief that there is a particular landscape human beings find attractive and delightful – a high clearing, with a wood behind it and a river and a prospect and a bluff. He says that this was a

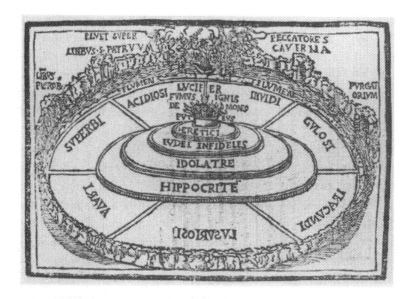

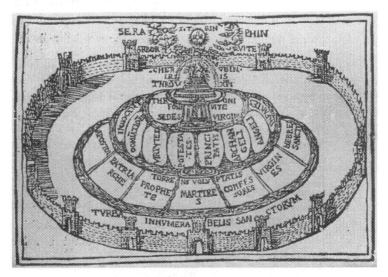

Figure 2 Memory theatres. These sixteenth-century Dante-esque diagrams illustrate how people used images as aids to memory. *Above*: Hell is divided into eleven places – steps for four types of sinner lead up to the central well, surrounded by locations for perpetrators of the seven deadly sins. *Below*: Paradise is surrounded by a wall sparkling with gems, and the throne of Christ lies in the middle.

A. S. Byatt

good place for primitive hunter–gatherers to feel safe and find food, but it is much the same landscape as the one described in John Armstrong's pursuit of 'The Paradise Myth' in paintings, poems and plays through the ages. Certainly by 'scanning' remembered constructed images of my own I can call up not only whole passages from many books and paintings and real places, but the connections between these, and projects for the future.

A concrete example of mnemonics within a text might be the use of repeated lists. I have spoken of the formal rhetorical device of the list of heroes, or places, in an epic – Milton's lists, in particular are delightful in the way they use the sensuous rhythms of the language and bring together, say, serpents from all histories and myths into one composite image. But lists work like the recapitulating pattern in a mnemonic. At the beginning of this chapter I quoted one such list, Proust's list of icons which solicited his own memory and desire to write: 'a cloud, a triangle, a belltower, a flower, a pebble'. It is rhetorically interesting to see how in the long passage at the end of *Le Temps Retrouvé* to which I have been referring throughout this chapter, he recapitulates repeatedly by listing the experiences which have formed, or will form, his novel.

> In fact . . . whether I was concerned with impressions like the one which I had received from the sight of the steeples of Martinville or with reminiscences like that of the unevenness of the two steps or the taste of the madeleine, the task was to interpret the given sensations as signs of so many laws and ideas, by trying to think – that is to say, to draw forth from the shadow – what I had merely felt, by trying to convert it into its spiritual equivalent. And this method, which seemed to me the sole method, what was it but the creation of a work of art? Already the consequences came flooding into my mind: first, whether I considered reminiscences of the kind evoked by the noise of the spoon or the taste of the madeleine, or those truths written by the aid of those shapes for whose meaning I searched in my brain, where – church steeples or wild grass growing in a wall – they composed a magical scrawl, complex and elaborate, their essential character was that I was not free to choose them, that such as they were they were given to me.
>
> (Proust, *op. cit.*, p. 232)

These lists, these recapitulations, give to Proust's long, winding sentences the form of memory and searching and connecting itself – the listed *things* and the *words* which are the symbols of these things, are the meshes of the net of my memory-metaphor, the knots which hold together and draw up objects in a coherent form. 'Coherent', or to use Coleridge's spelling, cohaerent, clinging together, is a word as important, in its different way, as 'haunting' when it comes to the vocabulary of memory in art. I ended *Sugar* with such a list, which has various functions. It is a recapitulation of the memories already elaborated in the preceding narrative – and of the ancient mnemonics which were brought into intricate life in the writing of the story. It is also an initiation of another cycle or recycle of the series of personal and family memories which it represents iconically. And it is mythical, I now see for the first time, in the *factual* way in which the Champs Elysées are mythical. It contains gold wings, Paradisal real apples and pears, and a furnace or oven through which one of my aunts, who gassed herself, entered the world of the dead. And it is an izibongo, an attempt to speak to my dead father. The moment described is that at which, as a child, I made a leap into my father's arms on the night of his unexpected return from the war.

> This event was a storied event, already lived over and over, in imagination and hope, in the invented future. The real thing, the true moment is as inaccessible as any point along that frantic leap. More things come back as I write; the gold-winged buttons on his jacket, forgotten between then and now. None of these words, none of these things, recall him. The gold-winged, fire-haired figure in the door is and was myth, though he did come back, he was there, at that time and I did make that leap. After things have happened, when we have taken a breath and a look, we begin to know what they are and were, we begin to tell them to ourselves. Fast, fast, these things took and take their place beside other markers, the teapot, the horse-trough, real apples and plums, a white ankle, a coalscuttle, two dolls in cellophane, a gas oven, a black and white dog, gold-winged buttons, the melded and twisting hanks of brown and white sugar.
>
> (Byatt, 1987, p. 248)

Proust compares his novel to a cathedral.

> In long books of this kind there are parts which there has been time only
> to sketch, parts which because of the very amplitude of the architect's
> plan, will no doubt never be completed. How many great cathedrals
> remain unfinished! The writer feeds his book, he strengthens the parts of
> it which are weak, he protects it, but afterwards it is the book that grows,
> that designates its author's tomb and defends it against the world's
> clamour and for a while against oblivion.
>
> (Proust, *op. cit.*, p. 431)

Frances Yates says,

> The high Gothic cathedral, as Panofsky has suggested, resembles a
> scholastic summa in being arranged according to a system of
> homologous parts and parts of parts. The extraordinary thought now
> arises that if Thomas Aquinas memorised his own *Summa* through
> 'corporeal similitudes' disposed on places following the order of its parts,
> the abstract *Summa* might be corporealised in memory into something
> like a Gothic cathedral full of images on its ordered places.
>
> (Yates, 1966, p. 79)

Both Proust's *roman-cathédrale*, and Yates's Summa-cathedral are
careful memorized constructs. Proust also has a different image of his
work – the novel as a dress, made of pieces tacked and fitted together, of
stuff, like a mosaic, to fit a body – or (he ends) like the different pieces of
[dead] jellied meat in Françoise's harmonized *boeuf en daube*. I see my
mnemonics as related to both cathedral and dress – except that I see the
dress as knitted, with the language as the yarn or thread, into which the
rhythmic pattern of stitches, colours and forms is worked.

Conclusion

I had meant to attach these writer's thoughts about memory to things I
have been reading about the biology of memory: I do feel that both
Proust's sense that whole memories (paradisal memories may be *triggered* by sensations) and my own sense of trawling a network for flashes
of significance (or preening hooks and eyes, *connections*) work well with
recent ideas of neural networks, with the discrete storage of parts of sensations and experiences in different parts of the brain, with the behaviour of axons and synapses as they are beginning to be understood.

Semir Zeki argues that the understanding of colour vision, for instance, has been delayed by the description of it as a dual process, in which a sensory 'impression' is then 'received' by an associational cortex. Colour vision, he says, involves what are thought of as 'higher' mental activities from its beginning – it involves memory, learning and judgement. Human beings, we are now told, differ from (most) other creatures because we can form images, in memory, of what was, but is not, there, and can use those images to make an idea of our continuous selves, a self-consciousness made of coherent memory images, which has what Hamlet said distinguished us from the beasts, 'discourse of reason, looking before and after'. We plan and imagine the future because we remember and form images of the past. Because we form images we are self-conscious – and with this reflexiveness, Gerald Edelman argues, came the fear of death, and of extinction of the self-conscious mind. Because we remember, because we form images, we are human, we name and record and compare colours. If I put it that way, I am almost able to answer the question I have asked myself ever since I realized that what I wanted to do was to make stories of life and death, to make images that will hold and place those glittering perceptions of which we appear to be *excessively*, superabundantly conscious. Why make works of art? Why not just live? Proust noted that the remembered experience was the paradisal one, *because it was a memory*. We are individuals because we are self-conscious and we are self-conscious because we make images. We are self-conscious and are therefore *interested* in the images we make and in the fact that we make them. We need to make images to try to understand the relation of our images to our lives and deaths. That is where art comes from.

I want to end with what I find the most moving image of both the mnemonic of the memory theatre, and of the making of images itself, Matisse's *Red Studio*, reproduced on the back of the jacket of this book. This has been called the greatest of Matisse's studio allegories, in which he assembles both the two-dimensional shadowy (but brilliantly dark red) world of his everyday surroundings, and a scattering of placed icons, recalled images of images, coherently connected by the simultaneous limitations and complexity of his palette. In some of these allegories the painter himself is present, a shadowy two-dimensional faceless figure, less real than the paintings; here there are only the paintings, so the paintings *are* the consciousness, as Proust became identical with his book.

I have stood for hours in front of this painting trying to follow the invisible threads which electrically link the different pinks and blues and greens. I counted fifteen pinks, once. It is an ambiguous picture, both flat and three-dimensional, and in that, I instinctively think, like many mnemonics, with its emphases and its background objects waiting to be scanned into significance. The clock is there, the drawers, the chair. The imaginary world lies brilliantly amongst these objects, which nevertheless map it and hold it together. It is paradisal brightness in the red cavern of the skull. Art is the subject of art, but this is because we are creatures that remember, make images, and we need to do this as much as we need to breathe and to feel our blood run. Art is life, in a way that earlier maps of mind and body did not quite understand.

FURTHER READING

Alexander, M. (ed.), *The Earliest English Poems*, London: Penguin Classics, 1966–77.

Armstrong, J., *The Paradise Myth*, Oxford: Oxford University Press, 1969.

Byatt, A. S., *Sugar and Other Stories*, London: Chatto & Windus, 1987.

Byatt, A. S., *Babel Tower*, London: Chatto & Windus, 1996.

Edelman, G., *Bright Air, Brilliant Fire on the Matter of the Mind*, London: Alan Lane, 1992. [See, for example, pp. 131–6 on higher-order consciousness and pp. 138–9 on desire for immortality and fear of death.]

MacDonald, R. R., *The Burial-Places of Memory: Epic Underworlds in Virgil, Dante and Milton*, Amherst: University of Massachusetts Press, 1987.

Opland, J., *Anglo-Saxon Oral Poetry: A Study of the Traditions*, London and New Haven, CN: Yale University Press, 1980.

Proust, M., *À la Recherche du Temps Perdu*, vol. 1, Paris: Bibliothèque de la Pléiade, Gallimard, 1954.

Proust, M., *À la Recherche du Temps Perdu* [In Search of Lost Time], transl. C. K. Scott Moncrieff and T. Kilmartin, revised by D. J. Enright, London: Chatto Edition, 1992; New York: Modern Library Edition, 1993.

Rorty, R., *Contingency, Irony and Solidarity*, Cambridge: Cambridge University Press, 1989.

Wilson, E. O., *Biophila*, Cambridge, MA: Harvard University Press, 1984.

Yates, F., *The Art of Memory*, Penguin: London, 1966.

Zeki, S., *A Vision of the Brain*, Oxford: Blackwell Scientific Publications, 1993.

4   Memory in Oral Tradition

JACK GOODY

Oral and literature cultures

In approaching this topic, I decided to start by discussing memory in oral cultures, which is what I call those without writing. Unlike many other scholars, I use the phrase 'oral tradition' to refer to what is transmitted orally in literate cultures. The two forms of oral transmission in societies with and without writing are often conflated, and that has been the case in the well-known work of Parry and Lord on the 'orality' of Homer. Most epics are products of literate cultures even if they are performed orally.

Oral performance in literate societies is undoubtedly influenced to different degrees by the presence of writing and should not be identified with the products of purely oral cultures. The point is not merely academic for it affects our understanding of much early literature and literary techniques, which are seen by many as marked by the so-called oral style. To push the point to a speculative level, speculative since I do not know a sufficient number of unwritten languages (and here translations are of no help whatsoever), many of the techniques we think of as oral seem to be rare in cultures without writing. Examples include assonance (as in *Beowulf* or the work of Gerard Manley Hopkins), mnemonic structure (as in the Sanskritic Rig-Veda), formulaic composition and even the very pervasive use of rhyme.

In his book *Primitive Song*, Maurice Bowra saw a metrical structure in such cultures as being required by the nature of the song, but argued there is 'nothing that can be called metre in the sense of having a regular number of strong beats, determined either by loudness or by the time it takes to pronounce a syllable'. Repetition is present but alliteration and rhyme, he writes, are incidental, if not quite accidental, ornaments, used

Jack Goody

very intermittently. Some writers agree that rhyme is rare in 'primitive literature', while others (Ruth Finnegan, for example) judge that 'it is probably among oral literatures in close contact with writing that full vowel and consonant rhyme is most significant. It occurs in late Latin songs, modern English nursery rhymes, and the rhymes of the troubadours, British ballads, Malay *pantun* quatrains and Irish political songs', as well as mediaeval Chinese ballads.

So these features are more characteristic of oral performance in literate cultures, where precise verbatim, word-for-word memorizing and recall become highly valued, rather than of the more flexible traditions I have encountered in purely oral cultures. Let me elaborate on this important difference.

Communication in oral cultures takes place overwhelmingly in face-to-face situations. Basically, information is stored in the memory, in the mind. Without writing there is virtually no storage of information outside the human brain and hence no distance communication over space and over time.

Plato decided that writing would ruin the memory and that oral man remembered much better than his literate counterparts. From Plato's time onwards we often hear tales of the phenomenal feats of memory in such cultures, of remembrancers who recalled the complexities of tribal histories, of bards who recited long myths or epics.

Those tales have continued to be told. In 1956, the American sociologist David Reisman claimed that members of oral cultures must have had good memories merely because nothing could be written down. That common assumption had even been made over twenty years earlier by Frederic Bartlett, author of a well-known book called *Remembering*, when he claimed that non-literate Africans had a special facility for low-level 'rote representation'. It is true that such societies are largely dependent on internal memory for transmitting culture, for the handing down of knowledge and customs from one generation to the next. However, they do not remember everything in a perfect form, that is, verbatim, by heart. I have known many people quite unable to give a consecutive account of the complex sequence of funeral or initiation rites. But, when the ceremonies actually start, one act leads to another until all is done; one person's recollection will help another. A sequence such as a burial also has a logic of its own. At times visual clues may jog the memory, as when an individual is finding his or her way from one place to another; as in a paperchase, one clue points to the rest.

However, other more critical observers have found rather poor verbatim memories in oral cultures. They attribute these to a lack of the general mnemonic skills and strategies that come with literacy and schooling, because non-literates performed badly in standard psychological memory experiments.

But what are these standard tests measuring? In psychologists' terms, memory in the tests means exact, verbatim recall. For example, you show somebody a list of objects, and ask him or her later to name the items, giving points for the right sequence or for simply remembering the items in any order. The process is like recalling the dates in history books.

But is that the type of memory that oral cultures really employ? It is true that some information has to be held in memory store in an exact (or more or less exact) form. I could not communicate with you by words unless we had similar lexicons. That is a prerequisite of social life, the basis of which is acquired early on in life. Children are mimetic. But what about adults, for example when learning lists of objects or words? I do not doubt there may be some occasions on which a member of an oral culture might find it useful to be able to recall items in this way (a list of trees for example) but from my own experience they are very few and far between. It is simply not a skill of any great value to people in oral cultures; that comes with schooling.

Storing knowledge

Many experiences are placed in memory store, but we do not need to keep most of these in a precise form. Indeed in oral cultures the notion of verbatim recall is hard to grasp. It is significant in that in the African languages with which I have worked I do not know a word for the memory (though there is a verb referring to action, which in LoDagaa is the word *tiera*, to think, or alternatively *bong*, to know; the concepts are not differentiated).

Everything that is learned goes into memory store in some form or other. At first this is by a process of imitation that enables us to acquire a particular language so that we can communicate in the cultural context in which we find ourselves. We may already possess the general in-built abilities to learn languages, but we need imitative learning to enable us to acquire exactly, precisely, verbatim, the basic features of a specific language. That imitative faculty, which seems to be an attribute of apes and humans but not of monkeys, is subject to fading; our ability to learn new languages fades rapidly after puberty. Young children acquire exact knowledge, verbatim, with an ease that disappears as we age; compare the capacity of young boys to remember the numbers of cars or engines with your own ability in this respect.

All cultural knowledge in oral cultures is stored in the mind, largely because there is little alternative. When we congratulate the members of oral cultures on their good memories, on one level we are simply saying that they have no other storage option. In contrast we can pull down books from our library shelves and consult them to find the reference for a quotation we vaguely remember, or to discover the name of a bird we did not know.

What is clear is that much oral knowledge is not stored in the precise way we think of in literate cultures when we talk of memory. In fact non-literates are often in the state of having a vague recollection in their heads, but of being unable to refer to a book in the way that we do. Hence they may just have to create new knowledge or new variants to fill the gap.

Mnemonic systems

Visual mnemonics provide some exceptions to this dependence on internal memory. However, mnemonics serve merely to jog the internal memory; they are no substitute for it, unlike a book or a computer. These mnemonics are material objects and sometimes graphic signs that fall short of fully fledged writing because they do not record linguistic expressions per se but only loosely refer to them.

We should be quite clear about the difference between a mnemonic system in oral cultures and the kind of recall that full writing permits and encourages. At one level they may be seen as equivalent, especially to Jacques Derrida and other post-modernists who are anxious to stress the relativity of cultures, and effectively deny hierarchical or evolutionary differences. But their position neglects cultural history, since culture does have a developmental history in which changes in the mode of communication are as important as those in the mode of production. For the information stored in mnemonic systems is rarely verbatim, word for word. Mnemonic systems present you with an object or a grapheme to remind you of an event or a recitation which you then elaborate. It is true that if I recite a short tale and offer you a mnemonic, you may be able to repeat it almost exactly. But such reproduction depends upon my telling you the tale and showing you the mnemonic. No one else can read it off directly from the object. There is no precise distance communication as there is with writing, which provides a fixed, society-wide code enabling us to establish more or less perfect linguistic communication over distance in time and space with people we may have never met but who have learnt the code.

You can do a lot with visual mnemonics. These precede writing, which is a code rather than a mnemonic, and are important for many purposes. But they have a very different role to play. Visual guides to territory (a kind of map) are said to be provided by the sacred *tchuringa* of the Australian Aborigines. These offer a new initiate a stylized picture of the local territory which he can then use in the walkabout he has to make at puberty; it introduces him not only to his local environment, both practically and symbolically, but also to kin and neighbours in the surrounding area.

Take another case, that of the Spokesman attached to the Asante court. Apart from knowing 'custom' in a general sense, he acts as a remembrancer in a more specific one. He is the one who knows the history of the stool, the throne, of the chiefdom, which he recounts on ceremonial occasions. Asante history is largely dynastic, and the Spokesman is reminded of this by the blackened 'stool' that each king leaves behind as an ancestral shrine to which sacrifices can be made (Figure 1). Associated with the stool may be various objects linked to events that have taken place during the king's reign. These objects, together with the marks on the shrine, including the remnants of the sacrifices that have been performed there, give the Spokesman mnemonic clues about the history itself. The drums may perform a similar role. 'Among the "insignia" of the Golden Stool of the Ashanti', writes Robert Rattray, in the 1920s, were 'iron and gold fetters, gold death-masks of great captains

Jack Goody

Figure 1 The Silver Stool of an Asante Queen Mother brought to England in 1874. The stool is itself a mnemonic.

and generals whom the Ashanti had slain in battle since the time of Osai Tutu', the founder of the dynasty. 'Among these were likenesses of Ntim Gyakari, King of Denkyira; Adinkira, King of Gyaman; Bra Kwante, King of Akyem; and Mankata'. The last name refers to Sir Charles M'Carthy, the British general killed by the Asante at the battle of Esamanko in 1824, whose skull now decorated the shrine.

Some of these histories are not so much memories as attempts to influence an audience about aspects of ownership of land or the extent of jurisdiction. In his enquiries in Asante, Rattray spoke of the great difficulty of getting at the truth. Even formal recitations are influenced by contemporary concerns. The divisional history recounted by Asumegya, begins with the first chief who came out of the ground on a Monday night followed by seven men, seven women, a leopard and a dog. The seven pairs (plus the chief of the Aduana clan) almost certainly represent the eight matrilineal clans or *abusua* that lie at the basis of Asante kinship and social organization. It is their present existence that prompted the history, not of course memory itself. In Asumegya's account, the first

Figure 2 Fomtomfrom drums from the Asante, round which one set human crania.

king was followed directly by one in whose time the Asante threw off the yoke of the Denkyira people and became independent; that is, it jumps to the seventeenth century. The doings of later kings in such histories may be recalled by some object tied to the stool which acts as an ancestor shrine or by a skull fastened to the state drums (Figure 2).

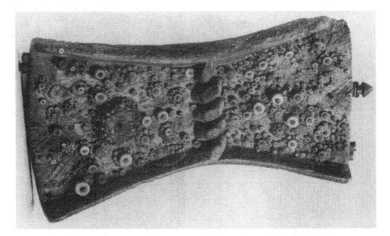

Figure 3 A divination or memory board (*lukasa*) from the Luba of Zaïre.

Among the Luba of the south-east of the Republic of Congo, genealogies are traced back to founding ancestors. While we may consider these as myth, the Luba see them as truthful; for them, present events are legitimized by their relationship to the sacred past, as enshrined in the charters for kingship. These charters were sacred, protected and disseminated by rigorously trained 'men of memory'. They could recite genealogies, lists of kings and the episodes in the kingship charter. During the eighteenth and nineteenth centuries, when the Luba kingdom was at its height, they created rituals for assisting memories and also invented mnemonic devices to assist historical recitation. Nevertheless the genealogies, like most genealogies, often reflected present interests rather than past facts.

Principal among these devices was the *lukasa*, a hand-held, wooden object studded with beads and pins, or covered with incised or carved symbols (Figure 3). In the catalogue of the 1995 exhibition of African art at the Royal Academy, edited by Tom Phillips, the boards are said 'to still serve as mnemonic devices, eliciting historical knowledge through their forms and iconography. Through oral narration and ritual performance, insignia serve both to conserve social values and to generate new values and interpretations of the past, as well as to effect general and political needs'. During Luba rituals to induct rulers into office, a *lukasa* is used to teach the sacred lore about culture heroes, clan migration, and the introduction of sacred rules; it also suggests the spatial positioning

Figure 4 James Red Sky, Senior, with his Ojibway birchbark scrolls.

of activities and offices within the kingdom or in a royal compound. Each *lukasa* elicits some or all of this information, but the narration varies with the knowledge and oratorical skill of the reader. That variation is critical.

Luba memory devices do not preserve thought or store language so much as stimulate them. They offered a multiplicity of meanings through their multi-referential iconography. Coloured beads, for example, refer to specific culture heroes, lines of beads to migrations. They also encrypt historical information about genealogies and lists of clans and kings. Yet the reading of these visual 'texts' varies from one occasion to the next, depending on the contingencies of local politics, and demonstrates that there is not an absolute or collective memory of Luba kingship but many memories and many histories.

Graphics, drawings which have made part of the journey to writing, may serve a similar purpose. In his account of the birchbark scrolls of the Ojibway of Canada, S. Dewdney explores the question of how in one particular form of shamanism, the teacher's version of a myth is prompted by a series of drawings. These map out, for example, the journey of a hero or the migration of a clan, and indicate the events that happened on the way (Figures 4, 5, and 6). These graphics are not writing in

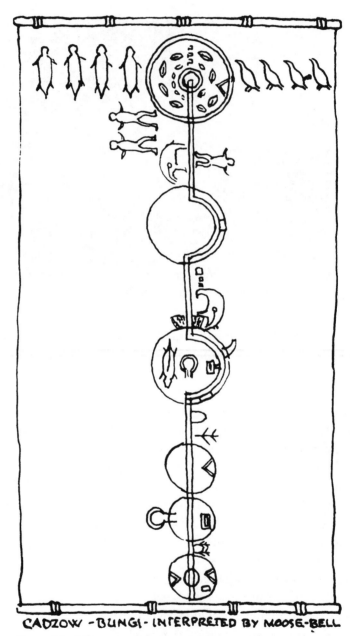

Figure 5 Red Sky's migration chart, the journey of the bear.

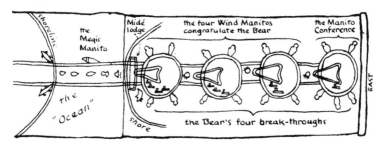

Figure 6 Scroll representing the first three degrees of Midé initiation, as interpreted by Moose Bell.

the full sense of the word. They serve as a mnemonic. No two interpreters will give altogether the same interpretation, even if they have learnt from the same master at the same time. This means that the divergence between one version and another is of an order quite different from that between two readers of the same linguistic text.

Memory and writing

I want to point up a paradox (perhaps a misconception) about memory in oral cultures and in the oral component of literate ones, the contrast I raised at the outset. In Jewish tradition, the word Mishna refers to the Oral Law, as opposed to Mikra, the written law. The Oral Law was delivered by God to Moses with the Pentateuch, but as an oral supplement. It now exists in written form but it is learnt by heart as we say, verbatim: it is in a sense preserved orally in memory, and if necessary may be transmitted by word of mouth. The word Mishna is in fact derived from the verb 'to teach', especially teaching by oral repetition. Indeed, 'the Mishna on Idolatory' has been described as apparently arranged with a view to facilitating the process of memorizing the contents. It appears that some digressions are really mnemonic aids based on the principle of association of ideas.

What is strange here is that at the very period in time when literacy made it possible to minimize memory storage, human society adopted the opposite tack, at least in some contexts. As W. Edwards, the editor of the Mishna, wrote in 1911, 'It is a point of curious interest to the modern mind that the great mass of knowledge and opinion . . . was retained in the schools for the most part, perhaps entirely, without the aid of writing. It was transmitted by memory alone, through the process of oral repetition, whence indeed came the title "Oral Tradition" '. But it is significant that the people in charge of this process were the Scribes, the Sopherim, the experts in writing themselves.

The role of the Scribes shows that this account is not wholly adequate. Students may hear a spoken version of a literate text and store that in memory. For example, many people learn the Lord's Prayer in this way but there is always a written version to check back upon. They do so because, perhaps following up Platonic fears, fully knowing a text meant learning it by heart. Some people, especially Christian fundamentalists, still set themselves this task for the whole Bible; more frequently it is done for the Qur'ān. Indeed Muslim schools are often directed to just this end; in non-Arabic speaking countries, the Book continues to be read and learnt in the original language (literally the words of God) which means that many simply know the text by heart without knowing Arabic or how to read it. If they also know the meaning, it is because they have learnt a parallel text (or possibly an utterance) at the same time; however, they cannot translate from one to the other or understand the text on its own. That form of learning leads to the development of verbatim memory skills. When these qualities were tested among the Vai of Liberia, Muslim literates, even if they could not understand Arabic, were more proficient than those literate in English or in Vai itself, or indeed in those who knew no script at all.

A similar situation seems to have occurred in the case of the Rig-Veda, a famous example for orientalists. Scholars often describe how this text has been transmitted orally from the senior generation of Brahmins to the next. Note that once again it is the scribes who are in charge of so-called oral transmission, in itself an indicative fact. For we know that the text had been available in written form since the fourteenth century, and indeed some scholars have suggested that it originated a thousand years earlier in 150 BCE (Before the Common Era) or even 500 BCE. Like the Mishna, the Rig-Veda contains many internal mnemonic devices; the tight phonetic structure of the verses seems to depend on alphabetic literacy and helps the learner recall exactly the content. It is always possible for the teacher to look back at the text and check to see that he or his pupils are not making a mistake. This form of transmission is associated with the Brahmanical tradition that the Vedas were *apanuruśeya*, produced by no human agency: they existed eternally and before the world itself. So for Brahmins there is no original written text composed either by God or by Man. Other Brahmanical textual traditions trace their origins to superhuman seers or *ŕśis*, who had, at least by implication, access to unconditioned texts that lend the tradition an aura of infallibility.

Similar considerations perhaps apply to the works of Homer, about which there has also been a long discussion as to whether they were composed orally in an oral culture. What is certain is that they were communicated by word of mouth, by recitation, to audiences at the annual festival of the Great Dionysia. On the other hand, they already existed in written form, since they were composed or transcribed around 650 BCE; their communication lay in the hands of Reciters (*rhapsodes*) rather than of lyric poets or composers (*aidos*).

Why should people who could read from the Bible, the Qur'ān, the Rig-Veda or Homer commit those works to memory and then produce them as spoken language? Partly because knowing a text, as with knowing a poem by Wordsworth, often means to memorize it word for word. 'Do you know "I wandered lonely as a Cloud"?' means 'Do you know it by heart?' You internalize the meaning with the words, which now become part of you, of your consciousness. To be able to quote in this way is prestigious – showing you are an educated man or woman as well as possibly being helpful to yourself in organizing your experience. To recall 'As flies to wanton boys are we to the Gods' may reinforce one's self-control under the blows of outrageous fortune. Furthermore, memorization is often necessary to turn a text into a performance (as with a play on stage), for broadcasting purposes. To speak the words of the gods is especially charismatic.

What difference does this make, the difference between oral communication as an internalized (written) text and as an utterance in a culture without writing? As members of a written culture we tend to read back our own memory procedures onto oral cultures. We look at oral cultures through literate eyes, whereas we need to look at orality from within.

Consider the difference. If I hear a longish poem or recitation in an oral culture, I can try to place it in my memory and later repeat it. The original reciter may even correct my repetition in an attempt to reproduce his or her own version. That procedure may lead one to assume that exact reproduction is the aim of this learning, a procedure that in this context allows no room for invention or creation. But leaving aside the intention of the actors, which is always uncertain, is it even theoretically possible to reproduce perfectly a recitation of any length in an oral culture?

In one famous set of experiments people in a circle were asked to pass a short verbal message from one individual to the next, unheard by the others. The message at the end was very different from that at the beginning. So the answer to the question is that exact reproduction of long recitations in purely oral cultures is very difficult indeed. When the teacher is absent, you have nothing to refer back to, no book, no model, except what remains in your memory. We know how fallacious that can be. Perhaps for reasons discussed by Freud, I may consistently transpose a name from Mr Brown to Mr Black. You know the sensation. You cannot stop yourself. If the teacher is absent and there is no fixed text, you pass on this change to all who hear you. The transmission between you and your audience is as it was between you and your teacher; that is, privileged. The recited version becomes the truth, becomes orthodoxy, and you become 'the authority'.

The Bagre myth

Let me give you a specific example of what happens in an oral culture with which I have worked for many years. I first recorded the Bagre myth of the LoDagaa of northern Ghana, which was part of the initiation rites of an association that provided medical and other benefits, in 1950. The association was secret and I was not a member. But a former resident of the village, called Benima, who had converted to Islam and now lived in a nearby strangers' settlement or *zongo*, understood my interest and offered to recite the myth to me as it had been taught to him by his father's brother. Over ten days, he recited while I wrote down his every word and got some partial explanations of the contents in a rough translation. I was partly convinced by what Benima said and others hinted at (and by the analysis of myth by many anthropologists such as Claude Levi-Strauss and Bronislaw Malinowski) that we were dealing with a fixed recitation that people knew by heart and that was handed down in more or less exact form, as a unique cultural expression, from generation to generation. Had this been true it would have been a fine example of exact memory in oral cultures. And indeed I was told how some new initiates who showed promise would be taught by an elder and their recital would be corrected by him.

I could not then easily test this assertion of fixity for it was before the days of portable taperecorders. But nevertheless I should have known better – partly because I knew about the circular whispering experiments, and partly too because, embedded within the Bagre that I had written down was an encouragement to neophytes to attend the performance of other groups and learn how they recited. In other words, variations were acknowledged and in a sense encouraged. The whole view of myth in oral cultures has suffered from the fact that until recently we usually had only one version of any long recitation (already a considerable achievement to record) which looked (and was discussed) as though it was a fixed, unique feature of that culture which should be analysed in a timeless way. If another version was recorded, the corpus was still treated as finite and variants were viewed as homologous alternatives of the same basic structure.

Some twenty years later, in the 1970s, I returned to that village as well as to neighbouring ones to make further recordings with my collaborator and co-author, Kum Gandah, who as a member of the association could attend the actual ceremonies. By now portable taperecorders had become available and we could record a number of versions on the spot and then transcribe and translate them later at our leisure. The whole task was a long one, spread over several years, but we completed some fifteen versions of the White Bagre (the first part of the recitation, closely tied to the ritual performances which acted as a mnemonic, a prompt) and nine of the Black Bagre (the second part, much more speculative, even philosophical, dealing with the acquisition of culture as well as with solutions to human troubles).

The differences were large (see Table 1). They were significant even when the same man recited on different occasions and greater still when different men recited on the same occasion (for the myth had to be recited three times at each ceremony). Between nearby settlements, only 16 kilometres apart, they are enormous. These differences have to do not only with length, i.e. whether some incidents have been included or

Table 1 *Length of recitation of the White Bagre*

		Number of lines
Benima, 1951, dictated		6133
Nyin, 1969, Bagre Bells		3917
Sielo, 1969, Bagre Bells (incomplete)		2778+
Nimidem and Sielo, 1969, Chief's House		3940
Nimidem, 1969		1704
Sielo, 1969		2236
Kob, 1969, Chief's House in the Black Bagre		2267
Sielo, 1974, open space in the Beating of the Malt		2239
Gbaa-Ziem, 1974, in the Beating of the Malt		1260
Sielo, 1974, at House in the Beating of the Malt		2225
Gomble, at the Bagre Dance		1204
Yikpɛɛ, 1st lineage, the Bagre Dance, Day	(a)	24
2, the morning of the Bagre morning	(b)	165
	(c)	84
Yikpɛɛ, 2nd lineage, the Bagre Dance	(a)	86
	(b)	114
	(c)	525

excluded. The differences are of a transformative, generative kind. Some stress the role of God as the creator of human culture; others emphasize that of the intermediaries, the beings of the wild (or fairies); yet others people's own contribution to their culture. Some are highly theocentric, others stress the role of other agencies. Nevertheless the LoDagaa will often assert the unity of the Bagre, that all versions are 'the same'. That is because all are performed in the course of the Bagre rites; it is like saying that the Anglican Evensong is the same whatever is included, different anthems, different prayers. But there is also another factor: when oral versions are recited at different times and places, you cannot easily make a comparison, not as you can when you lay the written versions of Evensong side by side and actually examine particular passages. In memory (as distinct from in archive) the versions tend to merge into one another.

What happens in the learning process is that the neophytes undergoing initiation in the Bagre sit patiently through these lengthy performances and those that show some talent as potential Speakers learn the mode of recitation, the *recitativo*, so they can fit any thought or sentence into the loose rhythmic structure of the verse. They listen, learn (without exactly memorizing), then, when called upon, give a performance of their own, retaining many incidents from those they have heard but omitting some and elaborating others.

Even what I first regarded as essential incidents are lost in other versions. In the first, dictated, version there was an account of one of the original humans crossing a river, which formed the boundary between this world and the next. Reaching the other side in a boat, he met with an old man smoking a pipe, who was God himself. As a result of God's invitation, the man climbed up to heaven with the aid of the Spider and his web which served as a ladder. There he was shown how God created Man. In the next version, which Gandah and I recorded some twenty years later, the whole of this incident, which at that time seemed to me structurally of the greatest importance, had been compressed into a few lines, which only I could now fill out. God played a very different role in the whole recitation.

So depending on which version you select, the analysis will vary greatly. And there is consequently much less close a fit between myth and other aspects of culture than many functionalists and structuralists have suggested. The indeterminancy is pervasive. There is a section at the beginning of each recitation consisting of some twelve lines which I have called the Invocation. It runs as follows:

The Invocation to the White Bagre

Gods,
ancestors,
guardians,
beings of the wild,
the leather bottles [which hold the divining cowries]
say we should perform,
because of the scorpion's sting,

because of suicide,
aches in the belly,
pains in the head.

Even when I had given up the idea that the Bagre was fixed, I still believed the Invocation to be rigid, because people would confidently begin to speak these lines, like reciting the Lord's Prayer. An elder would correct a younger man's version, and say, 'no not "hallowed" be *your* name" but "hallowed by *thy* name" '. However I have now recorded some dozen versions of these lines and none of them are precisely, word for word, the same as any other. If an elder corrects my recital, it is from his own memorized version, his personalized model, which differs slightly from that of others. Since there is no fixed text to correct from, variation is constantly creeping in, partly due to forgetting, partly due to perhaps unconscious attempts at improvement, at adjustment, at creation.

Replication and diversification

When you come to think of it, much cultural transmission has to be of this kind. It is a mistake to view the handing down of culture as the exact counterpart of genetic transmission, a kind of cultural mimesis; genetic reproduction is largely self-replicating, but human learning involves generative processes, what has been called 'learning to know'. Some changes in the corpus take place because of pressures from the outside, forcing or encouraging adjustment; others occur because of deliberately creative acts which have little to do with adjustments of this kind. That you can see by examining neighbouring cultures (as Frederik Barth does for New Guinea), cultures which presumably must have once been the same. The LoDagaa area consists of settlements and lineages each of which displays a number of differences from its neighbours. LoDagaa 'culture' is marked off only in a relative way from others, not by hard and fast boundaries. In fact customary behaviour changes, like the dialects of a language, gradually over a large region. Dialects can change only because memory, mimesis, imitation, is imperfect; forgetting may occur but in many cases perfect reproduction is not even aimed at. That is the same with other features of human cultures. In the absence of centralized mechanisms of control which inhibit change, differences in behaviour constantly emerge in this way; there is little of the precise replication involved in verbatim memory.

Oral memory is of course simply experience reworked. Only in the light of such a notion can we explain how neighbouring oral cultures, even very small ones, display these differences; the world offers us enormous diversity even in the simplest societies. Written cultures manage more easily to establish relatively stable internal norms, and in certain areas inhibit but do not prevent diversity. It is only in literate cultures such as those of Cambridge colleges that we find people reciting the same prayer every evening of the week, whatever the weather, whatever the national events, without any thought of innovation.

In other areas as well, writing encourages diversity, hence the high degree of turnover in the artistic world and the increased speed of the accumulation of knowledge. But in some situations writing establishes a conformity, an orthodoxy, as with Religions of the Book which endure unchanged over the centuries. Think for example of China, consisting of a fifth of the world's population. There one finds very similar marriage or funeral rites, stretching from the far north-east down to the south-west. Why? Because diversity of practice was deliberately suppressed by a civilizing process that introduced the Chinese script to tribal peoples. With that came books on ritual which prescribed exactly how one should act on these formal occasions. It was the duty of all citizens of the empire to follow these written rules and so to marry in a certain way. The written word sets up a situation which would be almost impossible in oral cultures. In the latter, storage is less exact, allowing more room for continuous creation of rituals but less for the cumulative accumulation of knowledge that literacy provides.

I do not mean to imply that in oral cultures people are incapable of exact reproduction from memory store. As we have seen, mimesis has to occur for children to be able to learn a language when there is little room for imprecise reproduction. That continues to be the case with other pieces of communication, short events such as proverbs and songs (where the rhythms help). It also occurs with short narrative sequences such as those found in folk-tales. These undergo less transformation than the Bagre (indeed, like Greek folk-tales in Homer, LoDagaa sequences are incorporated into the Bagre, where they constitute relatively constant elements in the recitations). Hence perhaps the broad similarity of many themes and shapes of folk-tales from culture to culture. But with longer oral forms, memorizing becomes more difficult and possibly less essential in the eyes of the people themselves.

As I have indicated, it is really with literacy and schooling that the task of having to recall exactly vast quantities of data (often half uncomprehendingly) is seen as necessary to keep culture going. It is interesting that the very earliest schools that taught writing, in Mesopotamia, required the pupils to read on one side of a clay tablet, then turn it over and write out what they had read on the other side. In other words, at the moment that writing had provided a storage system that did not require one to memorize, scholarship insisted upon exact verbatim memory, as if, anticipating Platonic fears, it distrusted what the new medium would do to people's mental processes.

For some purposes, such mimetic memorizing is necessary and advantageous. If you memorize the completely arbitrary order of the letters of the alphabet, which does not even group together vowels and consonants and which goes back to a Semitic proto-alphabet of 1500 BC, you have immense control over the information contained in a telephone annual, street directory or in the indices of books in any language using the Roman script. You will have more difficulty in Russia and be lost in Japan. But within the orbit of the same script you will manage to find the address and telephone number of a stranger in a strange town. To do this requires exact learning of a totally arbitrary kind. So too does much of the advancement of science. Remembering the Periodic Table or chemical formulae depends upon internalizing exactly less arbitrary forms of knowledge, since we may need to bring different elements of information together to solve a problem.

Conclusion

I have tried to refute a number of 'myths' about memory in oral cultures. Exact verbatim recall is in general more difficult than in written ones, partly because there is often no call for word-for-word repetition of a recitation like the Bagre. Every performance is also a creative act and there is no distinct separation between performer and creator; that dichotomy does not exist. For, *pace* Jacques Derrida, there is no archive in oral cultures of the same kind as our Public Record Office or the early libraries at Ebla or Alexandria. Of course a greater part of culture is carried in the mind, especially language, but not always in a fixed archival way; hence dialects proliferate. Like all cultures, oral ones depend on stored knowledge; however, much of this is stored in a way that cannot be recalled precisely, in the manner we usually relate to the psychological operation of memory. It is, as I have said, re-worked experience.

Marcel Jousse more or less identified memory with experience and human existence 'Un homme ne vaut que par ce qu'il a memorisé . . . La mémoire est intelligence approfondissante . . . La mémoire est tout l'homme et tout l'homme est mémoire'. ['A man is worth only what he has memorized . . . Memory is deepening intelligence . . . Memory is the whole of man and the whole of man is memory.' Original from G. Baron, *Mémoire Vivante: Vie et Oeuvre de Marcel Jousse*, 1981.] Jousse had a particular thesis in mind. For him Jesus was visualized as the Rabbi Iéshona of Nazareth who instructed his disciples in the Qehillà, 'l'assemblée mémorisante'. Using various techniques of what he called the oral style, the disciples memorized the words of the master in order to transmit them to the world at large.

The problem is that the so-called 'oral style' is a style of verbal discourse that characterizes a society with writing. The Rabbis are nothing if not literate; their work is to memorize the written text which they do by various mnemotechnical devices. In oral cultures, one is less concerned with the exact memory, which is important to Jousse: if the disciples learnt exactly by oral means then we may have the precise words of Jesus rather than a version reconstructed long afterwards. However, the mnemonic techniques which Jousse and others see as typical of oral discourse seem rather to be constructs of verbal communication in written cultures. That is possibly the case with strict poetic structures of versification and certainly with the kind of techniques used to memorize the Rig-Veda. These appear to be not so much residues of oral cultures but as the instruments of written ones. They are more frequently used in such a context because it is with literacy that the notion of exact verbatim reproduction becomes possible and valued. That is why it is associated with Brahmins and Rabbis.

Richard Sennett's chapter in this volume discusses collective memory, but it is dangerous to speak of a collective memory in oral cultures. An oral culture is not held in everybody's memory store (except for language, which largely is). Memories vary as does experience. Bits may be held by different people. They prompt one another regarding what to do at a marriage or a funeral and in so doing create cultural events anew. The notion of the static nature of oral cultures may, unlike ours, be partly true for the technological aspects. But for others there is greater flexibility. It is literate societies that proliferate mnemonic devices, not only in verse but of the spatial kind discussed by Frances Yates in her well-known book *The Art of Memory*. The limitations of memory in oral cultures, the role of forgetting and the generative use of language and gesture mean that human diversity is in a state of continuous creation, often cyclical rather than cumulative, even in the simplest of human societies.

FURTHER READING

Bartlett, F., *Remembering: A Study in Experimental and Social Psychology*, Cambridge: Cambridge University Press, 1932.

Dewdney, S., *The Sacred Scrolls of the Southern Ojibway*, Toronto: University of Toronto Press, 1975.

Finnegan, R., *Oral Poetry: Its Nature, Significance, and Social Context*, Cambridge: Cambridge University Press, 1977.

Goody, J., *The Myth of the Bagre*, Oxford: Clarendon Press, 1972.

Goody, J., *The Domestication of the Savage Mind*, Cambridge: Cambridge University Press, 1977.

Goody, J., *The Interface between the Written and the Oral*, Cambridge: Cambridge University Press, 1987.

Neisser, U., *Memory Observed: Remembering in Natural Contexts*, San Francisco: Freeman, 1932.

Yates, F., *The Art of Memory*, London: Routledge and Kegan Paul, 1966.

5 Memory and Psychoanalysis

JULIET MITCHELL

Although she came to see me as an analytical patient five times a week, Miss A. found it very difficult to remember anything from one session to the next. Shortly after we had met and started working together, she had told me that she needed me to remember why she had gone into a shoe shop. It was not like going into a grocer's and forgetting the sugar – anyone could do that; she did not even know why she was in the shoe shop. She added, somewhat embarrassed by what she had said, that she did not need me to know it too well, that would be absolutely awful.

Mr B., on the other hand, had an excellent memory and prided himself on recalling what I had said more accurately than I did. But one day when I mentioned something it turned out he had forgotten, he snapped at me, 'How can you expect me to remember *that* when I can't even remember my own mother from one day to the next?' Mr B.'s widowed mother needed his daily attention. Mr B. had a particular memory of childhood that was probably what we call a 'screen' memory – a mixture of experience and highly relevant fantasy, like a dream image. In this mnemic image he was standing shaking out a tarpaulin sheet together with his father, a DIY enthusiast: he could see the garden, himself, his hands on the tarpaulin, the tarpaulin held at both ends shaking, and then the image stopped. There was no-one he could call up in his mind's eye at the other end of the tarpaulin – a blank. This absence was covered over with what seemed to be a superb 'memory' for everything else.

At one of my many points of anxiety about writing this chapter I realized that, *sotte voce*, I had given myself a private subtitle which went: 'Psychoanalysis has progressed from understanding the amnesia of the hysterical girl as patient to the capacity to hold in mind an image of the psychoanalyst as mother.

Memory is an essential part of the process of humanization; psycho-analysis is concerned with the workings of, and the formation of, uncon-scious memory. It is not an easy subject.

At the end of the last century, when the enquiry that was to become psychoanalysis began, the first questions addressed the drastic gaps in memory, the 'absences', experienced by hysterical patients – typically young women – who presented in clinics and doctors' consulting rooms. Today in Britain the question of 'memory' within psychoanalysis has really come to rest with the task of the analyst to provide the context – the holding in mind, acting as 'container', offering a focused reverie – in which the baby/patient can come to have thoughts and memories of his or her own.

The British psychoanalyst Wilfred Bion theorized about how the mother's 'thinking ability', what he calls her 'alpha elements', could con-tain and process the raw material, the so-called 'beta elements', the undi-rected anxieties and sensations of the infant, and hand them back trans-formed into manageable feelings to the baby. Thereafter, the child should be able to use these elements or feelings for the formation of its own thinking and remembering. Miss A., in asking me to remember why she was in the shoe shop, was requesting that I do this for her – I should hold her overwhelming anxieties in my mind. But if I pushed my insights into her too forcefully, we later found out, they would become sexualized for her, penetrative and phallic, and arouse more anxieties than they resolved. We had to deal with her anxieties of course, but not in an intrusive way. Mr B. was telling me that, however well he could remem-ber details, this did not mean that he could hold either his mother or father in mind (his father was not there at the end of the tarpaulin). It frightened him to realize that his excellent memory for details had the effect of covering a hole; indeed perhaps that was, in part, its function.

The Freudian legacies

One of the earliest of Sigmund Freud's notions was that a neurotic symp-tom, especially an hysterical one, is the expression of a repressed memory. The symptom contains within it the representation of the agency that brought about the repression, and also the wish and impulse that had to be banished from consciousness but which has its own force

Figure 1 Sigmund Freud at Clark University, Worcester, Massachusetts, 1909.

and can reassert itself in this pathological, symptomatic form. This understanding of the formation of the symptom has not been abandoned; but the first therapy that emanated from it – recover the memory that has been repressed and the symptom will disappear – was not efficacious. If a treatment does not work, then the theory is either wrong in all or some respects or inadequate and must be refined and supplemented. Some tendencies within psychoanalysis have stayed more with the original understanding, others have emphasized supplements, others have moved to pastures new. To illustrate these three lines of development and also simply to try to control the arguments and theories about memory, I have isolated three very diverse strands representing distinct tendencies.

I am going to identify each with a geographical region. Such an account will be reductive; indeed, if it were not for the cohering force of a focus on memory, it would be misleading. All three tendencies can quite plausibly find their legitimation in aspects of Freud's work, some of which were contradictory aspects within that work, others of which were prevalent at one time and were then somewhat superceded but may have continued an 'underground' life.

I think different aspects of psychoanalytical understandings of memory are represented by:

(1) American ego-psychology.
(2) French structuralist, post-structuralist and deconstructionist thinking.
(3) British Object-Relations Theory.

In American ego-psychology, even where groups and individuals have broken away, the dominant paradigm concerning memory is the notion that experiences which are already constituted as potential memories are repressed. It is not by chance that psychohistory, the contemporary cult of narrative (everyone having their own story or history – 'what's your narrative?') and the Recovered Memory Movement all originate and expand in the psychodynamically informed culture of North America. Is this orientation to identity-as-history the product of a country that still feels itself a new nation, a nation in need of a past? In psychoanalytical treatment the patient must be helped to find his or her 'narrative'; the conflict-free ego must get hold of and make conscious, so as to take control of, the conflictual disturbances caused by the repressed memories in the unconscious and evidenced in symptoms or maladaptive behaviour.

For the French, memory is never constituted. It may seem odd to put such figures as Jacques Lacan, Jacques Derrida, Jean Laplanche and André Green into one camp. Even on the issue of memory, their creative work is often produced in disagreement with one another. When Lacan says that to read coffee grounds is not the same as to read hieroglyphics, one can imagine he is teasing Derrida for his notion of an arche-trace and the pre-writing of writing. Laplanche's emphasis on the 'enigmatic message' sent by the parent whom he nominates 'Other' (with a capital 'O') is a disputatious allusion to Lacan's Symbolic Other and Derrida's attack on Western logocentrism. All these issues are closely bound up with disputes over the question of memory.

Yet these diverse arguments and developments of ideas of memory all have a common base – a base very different from that which I have used America to represent. The key concept for the French is Freud's notion of a 'deferral' – *nachträglich, Nachträglichkeit* – first emphasized by Lacan, expanded by Derrida and deployed to develop his key concept of 'difference' and retranslated by Laplanche as 'afterwardness'. Memory comes into being only after the trace which marks it: there is no thing, no event, experience, feeling, to remember, there is only that present which an empty past brings into being. I will give a simple illustration – too simple, but something to hang on to. When it was still thought that sexual awareness arose only with puberty, Freud argued that any infantile experience, even a sexual one such as sexual abuse in early childhood, could be experienced as sexual only after puberty. The first – for the infant, non-sexual – experience (an experience empty of the sexuality that becomes its hallmark) is experienced as sexual later, in the present. It is not that the present reinterprets or gives meaning to the past, but that there is what we would call a retardation of meaning altogether from the viewpoint of memory – the past is nothing until it comes into being from the present. As Freud put it, 'What emerges from the unconscious is to be understood in the light not of what goes before but of what comes after.'

American ego-psychology, and Lacanian and post-Lacanian psychoanalysis, are usually virulently opposed to each other – or at least, ego-psychology was Lacan's *bête noire*. If we leave out Derrida for the moment, however, they have something in common which is, I believe, relevant to concepts of memory: they are orientated to language and to the father. In this respect British Object-Relations Theory is very different from both. Again the controversies within it are perhaps more important than its unity as an orientation – but that unity can represent a particular theory about memory. British psychoanalysis – whether it is Independent Object Relations or Kleinian or neo-Kleinian – sees itself as a two-person relationship in which the interaction between patient and analyst is the focal point. Although only words are used, these are an interpretation of a relationship which tends to be pre-verbal; affects and the body are used as a source of information about the psyche. Whether it is the reverie or facilitating environment that Donald Winnicott describes or the 'alpha function' of the analyst to contain and transform the anxieties of the patient, the model is of a mother: the nature of *her* memory is important for the development of memory in the infant. Instead of reconstructing a past like the Americans or deconstructing the past like the French, the British emphasize the so-called 'here and now' of the session. This 'here and now' dominates over any reconstruction of the patient's history through the patient's memory.

We have then, first, the notion that memories exist, are repressed and can be retrieved and that a history either of real experiences or of feelings and impulses can be reconstructed. Secondly, beside this or against this we have the thesis that memory is a series of inscriptions or traces which have no origin and no content in themselves. In a third perspective, the mother's/analyst's holding the baby in mind will facilitate the development of the pre-verbal baby's/patient's memory; but the relationship, not the memory, is what counts. These are not three psychoanalyses; they are three aspects of theories of memory which have received different degrees of emphasis. All three can find legitimation within Freud's work.

The significance of forgetting

What links these diverse strands is a focus on the *absence* of memory. Whether that absence is due to repression (the Americans), to deferral of meaning (the French), or to the immature developmental state of the patient as pre-verbal infant (the British), the starting point is absence. A hundred years ago psychoanalysis started not with memory but with forgetting. The pathological gaps in memory displayed by hysterical patients led Freud in time to observe and formulate a 'normal' universal amnesia of the first years of life: however hard we try, we do not remember our infancy in any continuous way. Between these two instances of forgetting – the hysterical pathological and the normal – (and really one could say because of them) came the great discoveries that are the objects of psychoanalytical theory and to a greater or lesser degree, depending on the psychoanalyst's orientation, the focal points of therapy: an unconscious which is structured and which functions in a way that is completely different from consciousness; repression and other modes of psychic defence; the Oedipus complex and infantile sexuality. One hundred years ago this was the field that was laid out between the hedgerows of the observation of hysterical forgetting and the theory of human infantile amnesia.

Hysterics seemed to be 'suffering from reminiscences' as well as having gaps in memory – as though they had both too much and too little memory. The reminiscences were cut-off bits of story somewhere between a daydream and a memory. The huge gaps were due to an internalized prohibition on thinking those thoughts or feeling those feelings. These feelings and thoughts were seen within the context of an event, so memories of these events had been 'repressed'. In 1910, Freud argued: 'All repressions are of *memories*, not of experiences, at most the latter are repressed in retrospect'. For reasons beyond the scope of this chapter, hysterical patients seemed to have repressed memories of sexual events, most particularly of intercourse with the father. When therapists of the Recovered Memory Movement, which has hit the USA like an epidemic, seek to release the dissociated or 'multiple personalities' of their clients from the defences they have erected, they search back to actual abusive events. Writers such as Frederick Crews have seen the feminist Recovered Memory therapists as Freud's 'true heirs'; according to Crews writing in the *New York Review of Books* (1995) 'the ties between Freud's methods (and theirs) are . . . intricate and enveloping – and immeasurably . . . compromising to both parties'. The Freud in question is the Freud who believed the stories that his first hysterical patients told when they were able to fill in the gaps.

My concern here is not with whether the stories were true or false. The result of coming to see them not as actual events but as fantasies was, quite simply, psychoanalysis. There would be no theory of the unconscious, of defences, of infantile sexuality or of the Oedipus complex if what we were dealing with were instances of specific abuse. There is no doubt that people who suffer trauma, at any age, have psychic responses. The importance of a shift from seeing hysterics as victims of individual acts of abuse to believing that human children, by virtue of their humanity, desire and repress a desire for their parents (the Oedipus complex) was that it changed the nature of the enquiry from one concerned with a discrete pathology to one emphasizing the formation of the human psyche. Recovered Memory therapists, by saying that patients with pathologies have suffered childhood abuse, are claiming that these people are not like the rest of us – they are a special population even if they exist in their hundreds of thousands. When Freud came to the conclusion that he, like his hysterical patients, was fantasizing his father's abuse of him in his childhood, he turned a marginalized, discrete pathology into a central aspect of the human condition. Psychoanalytical practice incorporates this shift: the analyst must first and in a sense, always, be a patient; neurosis and normalcy are on a continuum. Concepts of memory played a central part in this crucial change.

Following through from the very observable gaps in the memories of hysterical patients, Freud saw then that these were particular manifestations of a general human characteristic: we all forget our infancy. Though there is individual and possibly cultural variation, no-one has a continuous recall of the first years of life. Although there may be biological explanations, for psychoanalysts no physiological description can fully account for this observation. They feel that it can be explained by the particular nature of human interaction. In their view, the extreme dependence of the human infant induces a situation in which there is too much emotion, love and hate, towards he who protects and she who nourishes, and in the interests of human society, at a crucial stage of development, this excess must be forgotten, repressed. This act of repression makes the impulses unconscious and, because it is so major and momentous an act of obliteration, it drags with it all potential memories of this earliest period.

Although memory is individually and culturally quite variable, no psychoanalyst believes one can fully recover those earliest years of 'splendour in the grass' or of terror and anxiety as actual specific *memor-*

ies: the most is that they can be relived in the present of the therapeutic session so that something can be reconstructed.

One of the reasons why biology does not satisfactorily account for infantile amnesia is that there are some memories that seem to be very clear memories of childhood or even of infancy, memories that stand out from the general background of infantile amnesia with extraordinary clarity. One of my patients could clearly remember the first time she stood up. She had been placed on top of the fridge by her father who was holding her under the arms, and who then stood back to hold her just by her outstretched arms and fingertips. He let go so she stood for a moment, alone, ecstatic and shocked before she sat down with a bump on top of the fridge. The memory was incredibly vivid and various details made it possible to date it to somewhere around her ninth to eleventh month. This type of iconic memory is called a screen memory and, on analysis, would seem to be a mixture of a childhood experience and an unconscious fantasy. Its structure is like a symptom – something that has been repressed returns in a new, displaced image. If one can trace it to the fantasy and the experience then we have a clue to the infancy otherwise lost to amnesia. In 1914, Freud wrote of screen memories: 'Not only *some* but *all* of what is essential from childhood has been obtained in these memories. It is simply a question of knowing how to extract them out of analysis. They represent the forgotten years of childhood as adequately as the manifest content of a dream represents the dream-thoughts'. With the notion of 'screen memories' it is evident that we have a clear structure that indicates an underlying unconscious memory. Though in appearance like neither the symptom nor the dream (except in its clarity), its compromise formation signals it as a manifestation of the unconscious.

Can memories be false?

In his book, *Rewriting the Soul: Multiple Personality and the Sciences of Memory*, the Canadian philosopher Ian Hacking argues that in the twelve years from 1874 to 1886 in France, memory replaced the notion of the soul as the source and explanation of personal identity. This period saw the advent and efflorescence of the multiple personality.

Both similarly and differently from today's epidemic, recovering memory could produce the unified person who had split and dissociated to cope with trauma. Initially multiple personality was not a syndrome on its own, but was subsumed as a manifestation of hysteria. Hacking, a highly sophisticated, original and interesting thinker, wrote:

> The recovered memory people and the false memory people may seem completely at loggerheads, but they share a common assumption: either certain events occurred and were experienced, or they did not and were not. The past itself is determinate, true memory recalls these events as experienced, while a false one involves things that never happened. The objects to be remembered are definite and determinate, a reality prior to memory. *Even traditional psychoanalysis tends not to question the underlying definiteness of the past.* The analyst will be indifferent as to whether a recollected event really occurred. The present emotional meaning of the recollection is what counts. Nevertheless, the past itself, and how it was experienced at the time, is usually regarded as definite enough. [My italics.]
>
> (Hacking, 1995, p. 246)

Advocating the notion of the indeterminacy of the past, Hacking invokes the philosopher Elizabeth Anscombe's idea of an 'action under description' – a handshake can be goodbye, hello, clinching a business deal, a congratulation. In fact, despite what Hacking says, such indeterminacy is about as determinate as psychoanalytical concepts of the past have ever been.

In 1884 Freud studied with Jean-Martin Charcot at the Salpêtrière Hospital in Paris; he returned after a year to translate and introduce Charcot's ideas on hysteria to the Viennese medical community. Psycho-analysis can clearly be seen – and Hacking does thus see it – as emanating from within, and becoming exemplary of, what he designates 'the sciences of memory'. Discussions and theories of memory at the time were highly complex. Yet even when Freud was going like a sleuth after hysterics' apparent memories of incest, it was never with the notion of memory as a reproduction of a fixed event, true or untrue. The amalgamation of Recovered Memory therapy to psychoanalysis presupposes just such a notion to have been at work. There must be some explanation for the misunderstanding.

If we return to my initial distinction between American, French and

British concepts of memory within psychoanalysis, we can divide the subject, memory itself, in two. There are perceptions of experiences, whether internal or external to the subject, which follow old mnemic traces. Within this general category there is a second one of specific memories that are experiences or experiences of perceptions which are for some reason illicit sexual memories. Sexual memories become repressed and form part of what the psychoanalyst Melanie Klein called 'the repressed Unconscious'. It is to these which I believe many ego-psychologists address themselves. In one sense, because they follow old traces, it is as though these memories seem to be already constituted as memories and as such can be retrieved. Given this emphasis it becomes understandable (just) that non-psychoanalytical writers such as Crews and Hacking, despite the latter's sophistication, should place these memories in the same camp as those sought by the Recovered Memory therapists. Because what we might term these 'secondary memories' follow old tracks, it *appears* as if they originate from an actual starting point in childhood. In fact the trace but not the memory was there. These memories of illicit sexual scenes have been repressed by a process of secondary repression.

Even when he credited his own and his patients' memories of incest as truthful, Freud's notion of 'memory' was not one of literal reproduction. At this time, when he wrote to his friend Wilhelm Fliess, Freud was writing of memory in general, not just unconscious memories formed by the repression of sexual events or fantasies:

> I am working on the assumption that our psychological mechanism has come into being by a process of stratification: the material present in the form of memory-traces being subjected from time to time to a rearrangement in accordance with fresh circumstances – to a re-transcription. Thus what is essentially new about my theory is the thesis that memory is present not once but several times over, that it is laid down in various species of indications.
>
> (Freud, vol. 1, p. 233)

It is to this theory of the formation of memory as such that both the French and the British look, not to secondary repression of sexuality. Memories, then, are 'ideas' that flow over and over again along the same marks or traces. Consciousness is the state that is without such traces;

memory and consciousness are thus alternatives (they cannot happen at the same time). In his posthumously published *Project for a Scientific Psychology*, written in the late 1880s in the heyday of the sciences of memory, Freud tried to ground his psychological observations in neurology; although essentially he built on this earlier description of memory-making breaches in the psychic apparatus, later he used the image of a 'printator' or mystic writing pad to indicate how memories work. With a point one writes on a piece of cellophane that covers a sheet of wax paper, which together are attached at one end to a wax pad. One makes marks on the cellophane which go through to the pad beneath, but these marks can be wiped away by disturbing the contact between the cellophane and wax paper from the pad beneath. The cellophane is necessary to protect the vulnerable paper; we must have something that protects us from too many stimuli. In this way we can go on receiving impressions, recording them, and remain open to new ones. Memory is a process of forgetting, marking and being re-impressed.

Memories flow along already scored traces, the cellophane and wax paper representing the system of 'perception – consciousness'. They can be repeatedly cleared and made available for a re-inscription. However, if one examines the wax pad underneath, even when the paper is cleared the pad is scored over and over with a network of traces.

In the *Interpretation of Dreams* Freud describes the unconscious latent thoughts that one must hypothesize lie beneath the manifest thoughts that appear in the dream. These he said resemble the mycelium of the mushroom – there is no navel to the dream, no root, no origin or centre point, only a tangle of threads beneath the surface. We can transpose this image to memory: memory has no origin or root in a past object or experience: there are no memories *from* childhood, only *of* childhood.

The French and, very differently, the British interest is not with recovering secondarily repressed memories, but with the formation of memory itself. This formation of memory as such falls within a process known as primal repression. Primal repression is a necessary hypothesis: for something to be repressed at all it needs to be pushed into the unconscious from one direction and pulled into the unconscious by something already there that exerts an attraction. The question is, how can there be something that attracts which would have to have got there by the same process? A hypothetical explanation of the necessary existence of something already there, but without having an origin there, is given by Freud as follows:

> It is highly probable that the immediate precipitating causes of primal repression are quantitative factors such as an excessive degree of excitation and the breaking through of the protective shield against stimuli.
>
> (Freud, vol. 20, p. 94)

I shall come back to primal repression. The shift away from searching for the patient's memories necessitated also a shift away from the analyst's conscious memories towards a process of 'unconscious communication' between patient and analyst: 'It is a very remarkable thing that the unconscious of one human being can react upon that of another without passing through the conscious . . . [d]escriptively speaking, the fact is incontestable' (Freud, 1915, vol. 14, p. 194). I am going to give an instance to illustrate how this works and what relationship it bears to memory. This is really in order to by-pass giving an account of how psychoanalysis moved from eliciting the patient's memories to a formulation of the fundamental technique of 'free association'. The patient is asked to say whatever comes into his or her head, the aim being to escape the censorship that would otherwise operate to prohibit unconscious material from coming to the surface. Commensurate with this is the rule that the analyst should offer 'evenly suspended attention', listening with a part other than the logical conscious mind. Of unconscious communication, Freud wrote:

> Experience soon showed that the attitude which the analytic physician could most advantageously adopt was to surrender himself to his own unconscious mental activity, in a state of evenly suspended attention, to avoid as far as possible reflection and the construction of conscious expectations, not to try to fix anything that he heard particularly in his memory, and by these means to catch the drift of the patient's unconscious with his own unconscious.
>
> (Freud, vol. 14, p. 194)

Patients and analysts

In the following illustration, I am also hoping to show how this unconscious communication can open up a path into the patient's memories.

A patient, Mr D., has been complaining: he knows diagnosis is two-thirds of the problem – he corrects himself – he means two-thirds of the solution. But he finds that though he understands the situation better, he does not feel any better. After some expansion of this theme, he recounts a dream which he tells me has no bearing at all on what he has been saying. 'The dream is about something that is happening at work.' The Dream: *He edits an in-house magazine and it is due to go to press. Because he is ill the proofs had to be sent round to him at his house. [This really happened yesterday.] When he looks at them he discovers that although they had agreed what the finished product would be, this woman had just gone and changed the proof. He is very angry with her. Then he had to pay 65 000 dollars or it may have been pounds.*

Reconstructing later what goes through my mind, or rather what I hear at the time as though it were in italics or another print font – it is this: I heard the word 'proof' (used the second time without an 's') and in a very subterranean way in which I do not have time to linger, I remember he missed the word 'solution' and replaced it with 'problem' earlier. This is a new patient whom I do not know well, so somewhere I store the imagery of chemistry and wonder if he is scientifically trained or just a well-educated sceptic about the testability of psychoanalysis. The thought of alcohol goes through my mind. I notice I am confused by the figures he cites; I think I have heard '365' (pounds or dollars). I realize that during the thirty minutes of the session I have become confused as to whether at the very beginning of the session he had announced – as I think he did – that he has a fever now or that he had been telling me that he had one yesterday. As he starts talking about the Dream, I think he is having difficulty because the associations to the dream are not politically correct and he probably thinks of me as politically correct. I find myself wanting to reassure him that it is okay: we can all be angry with women who let us down. [I don't.] These are just some of the thoughts that come up when my conscious attention is suspended.

Laughing slightly apologetically, Mr D. tells me how this woman in the Dream is in fact a woman who works for him who has really betrayed him. She wants time off to have a baby and then wants to come back only part-time. In fact (it often happens) she may well not come back at all. The trouble is, he says, 'she is really good, so I feel *very* betrayed'. He talks about the fact that through the work with me he had understood more about his own situation and this had made him realize acutely that his wife couldn't stand any weakness. He wonders about the money in the dream. I do too, and noting the pounds or dollars, I make some suggestion about the fees of his previous American analyst and myself. My patient rejects this saying that he *had* thought about whether it had any bearing on my fees because last time he had dreamt numbers I had been right – the 300 that he had dreamt about did link directly to what he had paid me – but sixty-five did not link in any way. I said it sounded as though I had been pretty good at my job but had now betrayed him.

I then said I was confused: had he been ill yesterday or was he lying upon my couch today with a high temperature? He responded warmly that he was really ill now and felt dreadful but had wanted to keep his appointment. I said it certainly looked as though he felt we were both confused about what was the problem and what the solution: it was certainly no use diagnosing fever and then not being able to accept that someone who had fever would now feel weak. It was no use my helping him to see his situation if I then could not tolerate any weakness. My patient was visibly relieved by this but commented that his wife was working full-time and had the children so he could not expect her to be around all the time for him. I now decided to use what I had heard but what apparently he had not actually said: 365 where he had simply said 65. I would have left this alone as inexplicable or accidental if he had not followed my question about his American analyst (if it were dollars or pounds?) with the comment that I had been right previously when it had been a question of 300 but 65 did not make sense. It was as though the 300 in his mind had transmitted itself, through the communication of one unconscious to another, to my mind. So I now suggested that though he understood his wife had work and children and I had work and other patients, when he was feeling so ill and weak like a small child himself he would like us to be around 365 days of the year. At this, Mr D. started to talk about his mother's death in his early childhood and to bring up memories of how the nanny, who had had full-time care of him every day of the year, had left shortly afterwards . . .

Wilfred Bion said that as an analyst one must come to each session without either 'memory or desire'. This does not mean that I do not have my patient in mind, but that if I consciously recall last week's session or his childhood or prompt him to have memories it will interfere with the unconscious communication between his free-floating association and my suspension of my conscious attention.

When psychoanalysis came into being at the turn of the last century a prevalent explanation of pathological symptoms, above all hysterical, was an earlier trauma. This was almost always given as the background to the male hysteria that Charcot found in his clinic in Paris where Freud worked. For a long time Freud – like other clinicians – subscribed to the aetiological importance of trauma, but its significance faded with the notion of the power of wishes and their prohibition. Trauma came into focus again in World War I when shell-shocked men produced the same symptoms as had hysterical women with their unresolved Oedipus complexes. The analyst Otto Rank advanced the thesis that neurosis was caused not by repressed sexual desires as Freud had established, but by the trauma of birth. Freud disagreed, but changed his position and subsequently argued that, though there could be anxiety from repressed sexuality, there was also a primary anxiety in the infant. Traumas are also compulsively relived: the notion of a death drive (a force in humankind pushing backwards towards inertia or the inorganic) is hypothesized to account for this.

The Recovered Memory therapists (see for example Judith Lewis Herman's *Trauma and Recovery*) have put trauma on centre stage, in fact have amalgamated all pathologies to expressions of trauma. When one works as a clinician, whatever one thinks about is, as it were, at first 'involuntarily' thought through one's patients. When I did a head count of the patients who had come most to mind (but were not consciously selected) for this talk on memory – all had had childhood traumas, not of abuse but of loss or separation severe enough to be experienced as death. Miss A. who needed me to remember why she had gone into a shoe shop seemed to have received most affection from her father but had then been separated from him between the ages of about two and six – more than long enough to have no conscious memory of him whatsoever. Mr B. who had no image at the end of the sheet of tarpaulin and could not hold in mind his widowed mother, whom he visited daily, had been bereaved of his father in childhood. Miss C.'s screen memory *did* –

as screen memories do – encapsulate her infancy: her mother had died of a septic abortion when Miss C. was six months old and her father had cared for her until she was about three, when he had left her in foster-care while he moved away to form a new family. Excited and terrified at standing on her own two feet with her father's support (under the arms and then just by the fingertips), she had collapsed psychically – as she had collapsed on the fridge top – just after he left her for good. Mr D.'s mother had died and his nanny had left shortly afterwards; because there were two of them, his wife and I were always paired for Mr D.

Memory of course is no less relevant for all one's patients who have not had traumas – indeed for everyone. I was just curious that, with no conscious planning on my part these were the patients that came most to mind; only afterwards did I realize what they had in common.

Twice over – in the 1890s and after World War I – after giving it great credence, Freud rejected the possibility that although, of course, people have a psychic response to trauma, trauma itself could be the explanation of neurosis. However, in psychoanalysis, as in the formation of memory itself, ideas that are entertained rarely disappear altogether. If we look back at the theories of memory, or the hypothesis of primal repression which establishes the fundament of unconscious memories, what are these theories, if not modelled on trauma?

Psychic trauma like physical trauma breaks through the subject's pro-tective shield so that there is an influx of excitations which cannot be mastered or tolerated; according to Freud, 'primal repression' comes about with an 'excessive degree of excitation and the breaking through of the protective shield against stimuli'. The mystic pad, as a model for memory, describes a breaching of a protective cellophane and wax paper to form the ineradicable, permanent marks below; always the lan-guage is of quantities of excitation and of breaching. Derrida, in *Writing and Difference* (1978), glosses Freud's ideas on the formation of memory thus: 'Life already threatened by the origin of memory which constitutes it, and by the breaching which it resists, the effraction which it can con-tain only by repeating it'.

Compared to other animals, human beings are born prematurely, eyes wide open; after the first week the eyes become less focused. It is

Juliet Mitchell

common for actually traumatized babies to stare longer and harder and
then to retreat further into infantile autism. But internally and externally
there is too much for any human neonate to take in all at once. The neo-
teny of this early birth is a condition of potential trauma. Through primal
repression the first shock of too much excitation forms an unconscious
part of the personality with the help of the first care-giver and the mutu-
ality of language – most often, the mother matches words to infant's
sounds and infant matches sounds to mother's words. Later, the isolated
images can become connected and social, or they can be taken in or be
pushed in and stored as memories, the mixtures of feelings in the context
of experiences. But because the world is always too much with us, memories can never be replications nor even the same as themselves, not
from one day or one minute to the next.

The British Object-Relations school of psychoanalysis illustrates how,
without the other of (m)other or analyst, the helpless neonate could not
develop a system of memory; as well as gaps in memory, there is too
much consciousness in the traumatized child. Without the capacity for
memory which acts as a safeguard, a protection against the potential
trauma of being helpless in the face of the superabundance of the world,
there could be no human society.

FURTHER READING
Freud, S., *Psychoanalysis and Faith. The Letters of Sigmund Freud to Oscar Pfister*,
    transl. E. Mosbacher, London: The Hogarth Press and Institute of
    Psychoanalysis, 1963.
Hacking, I., *Rewriting the Soul: Multiple Personality and the Science of Memory*,
    Princeton, NJ: Princeton University Press, 1995.
Strachey, J. (ed. and transl.), *The Standard Edition of the Complete Psychological
    Works of Sigmund Freud*, vols. 1–24, London: The Hogarth Press and the
    Institute of Psychoanalysis, 1953–74, reprinted 1981.

112

6   When Memory Fails

BARBARA A. WILSON

Although all of us experience memory failures at some time or another, these slips of memory do not cause severe disruptions to our daily lives. Most of us can still function adequately at work, engage in conversation, and remember the gist of the programme we saw on television last night, while accepting as normal the forgetting of certain details. After all, nobody remembers *everything*. For some people, however, their memory failure is of such a proportion that the effects can be devastating.

Imagine waking up and not being able to remember what you did yesterday. Imagine living in a time vacuum where there is no past to anchor the present and no future to anticipate. Such is the fate of many people suffering from *organic amnesia*. Although amnesia means literally 'an absence of memory', in practice people with organic amnesia do not have a total loss of memory. They remember who they are, they remember how to talk, and how to read, and they usually remember how to do things they learned before the onset of their memory loss, such as how to swim, ride a bike or play the piano. Unfortunately, they have great difficulty in learning new skills or information, experience problems when trying to remember ongoing events, and usually have a memory gap for the few days, weeks, months or even years before becoming ill.

In contrast, people with *functional amnesia* following, say, an emotional trauma, sometimes seem to lose memory for personal identity. Despite the rather exaggerated interest in functional amnesia from the media, organic amnesia is far more commonly encountered in clinical practice.

People with organic memory problems whose other cognitive skills remain unimpaired are considered to have a *pure amnesic syndrome*.

These cases are rare in comparison with those where memory is only one of several cognitive impairments, such as word-finding difficulties, attention problems, and slowed thinking or processing of information. People suffering from these more generalized cognitive impairments are likely to be described as suffering from *memory impairment* rather than amnesia.

Memory problems are common after any brain disease or injury, and include such degenerative conditions as Alzheimer's disease or multiple sclerosis, traumatic head injury, stroke, encephalitis (a viral infection of the brain), hypoxic brain damage (caused by shortage of oxygen to the brain), cerebral tumour and Korsakov's syndrome (a condition caused by chronic alcohol abuse and poor nutrition).

Although it is not possible to restore lost memory functioning once the period of natural recovery has passed, it is nevertheless possible to help memory-impaired people compensate for or by-pass some of their problems and thus lead a more normal life than might have been possible without treatment. This essay presents a number of case studies of people with memory problems resulting from some of the causes described above: it will show how memory problems have affected the lives of these people, and discuss ways in which these individuals and their families have been taught to cope with problems occurring in their daily lives.

Before turning to individual cases, however, it is necessary to consider the brain regions that are most critical for establishing new memories and maintaining old ones. Furthermore, we need a more precise description of the nature and subcomponent processes of the multifaceted cognitive domain called memory.

Basic anatomy

One reason why memory is vulnerable to so many neurological conditions is that several structures in the brain appear to be critical. Consequently, if any of these structures, or the connections between them, are damaged, memory functioning is impaired. The two main anatomical regions involved are the frontal and temporal lobes, both of which are likely to be damaged following severe traumatic head injury. Beneath these cortical regions are subcortical structures, some of which are also

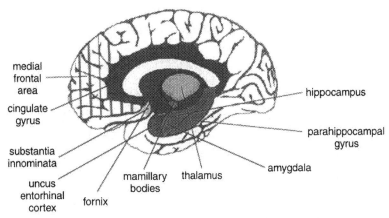

Figure 1 Subcortical structures involved in memory functioning.

crucial to adequate memory functioning. Damage to the hippocampus, the amygdala, the fornix and the thalamus, for example, are involved in many cases of severe amnesia. These structures can be seen in Figure 1.

Different kinds of memory

We often talk about memory as if it were one skill or one function. We say things like, 'I have a terrible memory' or 'She has a photographic memory', when in fact memory comprises a number of subskills, subsystems or subfunctions working together. The number of these subdivisions depends upon the model or classification system to which one adheres. An influential model proposed by Alan Baddeley and Graham Hitch in 1974 is one that subdivides memory into three main types, depending both on time-based and conceptual differences characterizing the types. This model can be seen in Figure 2.

(1) *Sensory memory* – the brief and rather literal trace that results from a visual, auditory or other sensory event – probably lasts only for about a quarter of a second.

(2) *Working memory* is considered to have several main components or functions. The first, short-term or immediate memory, holds recently arrived information for several seconds or even minutes, so long as the person is focusing on or 'rehearsing' the information. Unlike sensory memory, information in working memory has already undergone substantial cognitive analysis, so that it is typically represented in meaningful chunks; for example, words. The second is a central

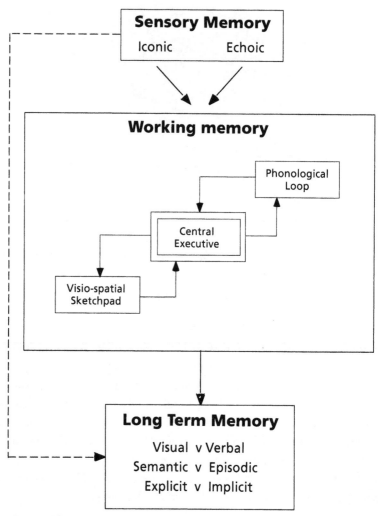

Figure 2 The Working Memory Model proposed by Alan Baddeley and Graham Hitch.

executive that can be conceived of as a controller or organizer and allocator of resources.

(3) *Long-term memory* is the system that encodes information in a reasonably robust (though not necessarily literal) form which can last for decades.

Although the very brief sensory memory system can in theory be subdivided into stores for all five senses, in practice only the visual and auditory sensory memory systems have been studied in any detail. Visual or 'iconic' sensory memory is the system involved, for example, in watching a film: our experience is of a continuous, moving film when in fact we are seeing a series of still shots presented very rapidly one after the other. Because each shot is held in this very brief visual sensory store, we perceive the film as moving. Patients with deficits in this system present as having perceptual problems, and thus we would not normally think of them as memory impaired.

Similarly, the auditory or 'echoic' sensory memory system serves the perception and understanding of language. Consequently, patients with auditory sensory memory deficits will present as having language comprehension problems.

At the centre of the working memory section of the model is the central executive, whose function is to enable us to plan and organize our attentional and cognitive resources. Suppose you are driving along a familiar road with a passenger. You are perfectly able to drive and talk at the same time: that is, you can allocate sufficient resources to accomplish both tasks. However, suppose that something unexpected happens on the road and you have to take evasive action. You will probably stop talking to your passenger because you now need all your attention for the task of driving. It is the central executive that enables you to reallocate your cognitive resources in such circumstances. Patients with frontal lobe damage often have problems with tasks requiring a shifting of resources in complex and demanding situations. They have a deficit of the central executive, and although able to work well at routine or automatic tasks, have problems in novel or non-routine situations. Once again, however, this is not exactly what one would think of as a memory deficit per se but rather a deficit of planning, organization and divided attention.

The central executive is supported by subsidiary systems, two of which have been studied in some detail. These are the phonological loop and the visuo-spatial sketch pad. The phonological loop is the system we employ, for example, if we look up or hear a new telephone number and then need to retain the seven or so digits long enough to dial the number. If the number is engaged or if someone talks to us when we are dialling, the number is vulnerable to forgetting and we often have to look it up

Barbara A. Wilson

(a)

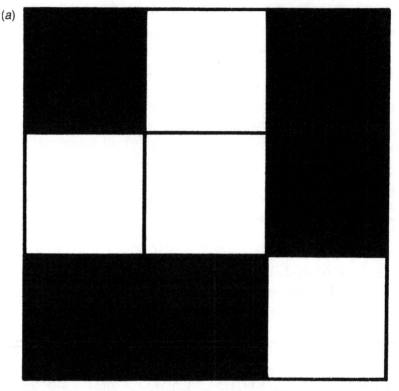

Figure 3 An example of a visual immediate memory task. (*a*) is shown for 5 s then removed and followed by (*b*). The subject has to indicate which square is different in (*b*).

again. The phonological loop also enables us to manipulate phonological information: if, for example, you were asked to delete the initial sound of a word such as <u>ch</u>in, and place the sound at the end of the word, making in<u>ch</u>, you would need your phonological loop to hold the sound of 'chin' (and the instruction) while moving the required phonological segment. Patients with phonological loop deficits find such tasks difficult or impossible.

Phonological loop deficits are seen mostly in the context of more widespread language impairments, although it is sometimes possible to find such patients with relatively normal language skills as measured by naming and speaking. These patients have a short-term or immediate memory deficit: if asked, for example, to repeat a series of digits in the same order as spoken by someone else, they would probably be able to

(b)

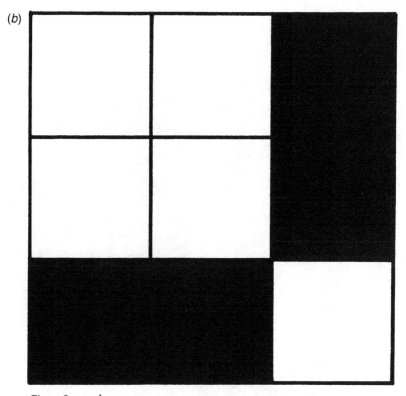

Figure 3 contd

repeat only one or two correctly. Most cognitively normal people can reproduce a sequence of around seven digits in the correct order if recall immediately follows presentation.

People with immediate memory deficits will also have difficulty learning new vocabulary or foreign languages. They may have problems with long sentences and may be unable to complete mental arithmetic operations because they can keep only a few items in mind at any one time. Nevertheless, they are not representative of memory-impaired patients in general, who typically have problems after a delay or distraction.

The visuo-spatial sketchpad is the non-verbal equivalent of the phonological loop. If you were asked how many windows there are in your house, you would probably call up an image of your house and mentally walk round the rooms counting the windows; this requires the visuo-spatial sketch pad. Patients with deficits in this system are unable to call

Barbara A. Wilson

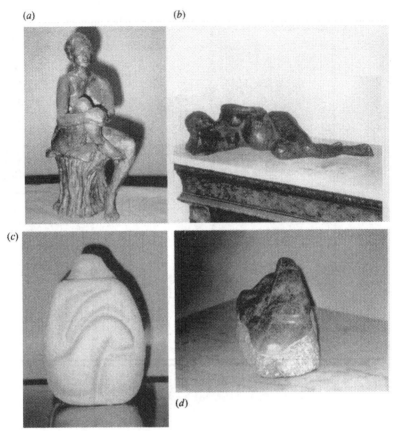

Figure 4 Sculptures before (*a, b*) and after (*c, d*) a neurological insult to the sculptress.

up images easily and also have difficulty with visual immediate memory tasks. So, for example, if they were shown a matrix such as that in Figure 3a for five seconds, then immediately afterwards shown another matrix such as that in Figure 3b, they would find it difficult or even impossible to indicate which square was different.

A sculptress I saw recently had great difficulty when faced with this kind of task. Her deficit in this aspect of working memory affected her sculpture: before her neurological damage, her figures were sculpted in considerable natural detail (see Figures 4a and b), but afterwards they had a much more abstract character, without delineation of detail (see Figures 4c and d).

120

Once again I must stress that the problems I have just described are not typical of those seen in the majority of memory-impaired patients, most of whom have problems with *long-term memory*. Patients with an organic amnesic syndrome have:

(1) profound difficulty in learning and remembering most kinds of new information (*anterograde amnesia*);
(2) difficulty remembering some information acquired before the onset of the brain injury or disease (*retrograde amnesia*);
(3) relatively normal immediate (working) memory; and
(4) relatively normal learning on at least some tasks of implicit memory; that is, learning without awareness, or where conscious recollection of the original learning is not essential for performance.

These characteristics are seen in people with a pure organic amnesic syndrome, who may have good cognitive functioning apart from their disabled memory. However, most memory-impaired people *do* have additional cognitive deficits, which are manifested in problems of attention, word finding and reasoning that make them slower at thinking and processing information. Rehabilitation in these cases is made more complicated and difficult because of the multiplicity of cognitive deficits.

Whether the amnesia is pure or not, a clinician or therapist working with memory-impaired patients will typically see people who have:

(1) reasonable immediate memory;
(2) difficulty remembering after a delay or distraction;
(3) difficulty learning most new information;
(4) better recall of events that happened a long time before the brain injury or disease than of events just preceding it.

Organic or functional amnesia?

So far I have discussed organic amnesia, which is amnesia caused by injury to or disease of the brain. Despite the frequently presented view in the media that amnesia follows from emotional trauma, functional or emotional amnesia is, as I mentioned earlier, far less common than organic amnesia. How does organic amnesia differ from functional amnesia? It is not always a clear-cut distinction, and indeed some people have memory problems resulting both from brain injury and from an overlying or functional memory disorder, possibly caused by a need for sympathy or some other secondary gain. Nevertheless, when asked to determine whether a person has an organic memory disorder, neuropsychologists and neurologists rely on several specific factors. The most prominent of these is the presence of clear physical damage to the brain. In modern neurology, several methods of scanning or imaging the brain – CT (computed tomography) and MRI (magnetic resonance imaging) – typically reveal severe structural damage to the memory areas of the brain in patients with organic amnesia. Such structural abnormalities are absent or minimal in functional amnesia.

As mentioned above, most people with organic memory problems have normal or nearly normal scores on immediate memory tasks, whereas some people with functional amnesia are poor at both immediate and delayed memory tasks. Furthermore, people with organic memory problems typically show an exaggeration of the normal advantage on tests merely requiring recognition memory as compared with the more difficult task of recall; but people with functional amnesia may actually show a reversal of this pattern. There are certain tests available which, although they look demanding, are in fact relatively easy for people with organic memory problems. However, people with functional amnesia sometimes score poorly on these tests.

So it follows that a number of things have to be taken into account when trying to decide whether a particular memory impairment can be diagnosed as organic or functional. In most cases of organic disorder, however, the behavioural manifestations can be linked to a clear physical cause: a patient has sustained an insult to the brain from one of various sources. Sudden onset of brain injury results from either severe traumatic head injury (as in many road traffic accidents), or cerebrovascular accident (stroke), or loss of oxygen supply to the brain, or viral encephalitis. Gradual erosion of memory function results from neurodegenerative illnesses such as Alzheimer's disease. Only in a very small minority of cases will there be a suspicion that the amnesia might be functional in nature.

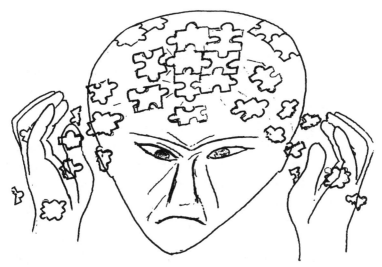

Figure 5 Drawing by a woman with encephalitis: 'What it feels like to be amnesic'.

Feelings and emotions of people with organic memory impairment

Figure 5 shows a drawing completed by a woman with memory problems (following encephalitis) who wanted to show in pictorial form how she felt about her difficulties. Although severe memory impairment is always extremely distressing for family members and carers, these feelings of distress are not always shared by the actual person with the impairment. Usually, lack of concern on the patient's part is associated with the poor insight that can follow from the original injury. Not surprisingly, those with greater awareness of their difficulties are typically distressed and/or depressed. They may feel that their memory disorder looks like stupidity in the eyes of others, or they may believe that they are going crazy, or that other people will regard them as mentally unstable. Some memory-impaired people report that their life is like a dream, and one young woman did in fact repeat to me several times each day the words 'This is a dream, I am going to wake up.' Some people, aware that they may have repeated the same story over and over again, become very withdrawn and socially isolated because of their reluctance to bore and irritate their families and friends.

Amnesia can also be exhausting for both the relatives and the memory-impaired person, because of the efforts required to cope with daily living and the compensatory behaviour necessary to get through each day. People with amnesia are also frustrated by frequent and sometimes humiliating failures and constant misunderstandings. As one young man said, 'Frustration is constantly waiting in the wings.'

Another young man became amnesic after a suicide attempt with carbon monoxide when he was nineteen years old. Following rehabilitation, Jack learned to compensate reasonably well for his poor memory and eventually went to college where he enrolled in a media studies course. Jack's problems were explained to the teaching staff who, by and large, organized his timetable appropriately and made allowances for his impairments. Things, however, did not always go smoothly. For example, Jack was once asked to relay a message by a college lecturer. Afterwards, he recounted: 'I left the room and promptly forgot the message, where to go, the name of the lecturer who sent me, and how to return to the room I had been sent from. I was lost and I can't recall what became of the incident. But the feelings of indignation, frustration and fear stayed with me.' Jack also went on to write, 'I might be mistaken in this. I cannot gauge how accurate any of my memories are.'

Amnesia: the two extremes

First, in this section, I shall describe C.W., a man with extremely severe memory problems for whom efforts to circumvent his memory difficulties have proved almost wholly ineffectual. C.W.'s plight will be contrasted with the relative success of J.C., a young man who has learned to compensate well for his memory deficits. Reasons for their contrasting outcomes will become evident to the reader as the text continues.

C.W. developed herpes simplex encephalitis (a viral infection that can sometimes invade brain structures) in 1985 when he was forty-six years old. An outstanding musician, C.W. was a producer of early music for national radio, chorus master of a leading contemporary music group, conductor of a professional Renaissance group, and one of the world's leading authorities on Orlando Lassus, a Renaissance composer.

A computerized scan of C.W.'s brain soon after his illness showed severe damage to both temporal lobes and to part of the left frontal lobe.

A more detailed and sophisticated magnetic resonance imaging scan, carried out in 1991, showed both temporal lobes to be abnormal: the hippocampal region was affected bilaterally, with the left being completely destroyed; both mamillary bodies were affected and there was further slight damage to the left medial frontal region. The MRI scan can be seen in Figure 6.

I first saw C.W. in October 1985, some seven months after the onset of his illness. He was not an easy man to assess, partly because he reported every few minutes that he had just woken up. This insistence on C.W.'s part that he has just woken and is experiencing the world as it were for the first time has continued for eleven years. He tries to capture the moment of just awakening in his diary, an example of which can be seen in Figure 7. Entries to the diary such as these occur many times each day, seven days a week, fifty-two weeks a year.

After considerable time and effort, C.W.'s wife and health-care staff found ways of by-passing some of his problems by changing their own behaviour. For example, when C.W. states that he has just woken up, sympathizing with him or attempting to offer an explanation appears to increase both his agitation and the rate of repetition of the phrase. Changing the subject or distracting him, on the other hand, reduces his agitation and repetition frequency. Because C.W. is unable to remember people involved in his care, he considers them to be strangers, and consequently it is necessary to treat him formally. Rather than use his first name, which staff would normally do with patients whom they know for many years, they refer to him as Mr W. Although this may seem a small detail of treatment, changing the behaviour of those interacting with memory-impaired people can be an effective way of reducing some problems for the amnesic person. It can be regarded as a form of environmental change that avoids the need to remember.

In contrast to C.W. is J.C., a young man who became amnesic at the age of twenty years following a brain haemorrhage that damaged the hippocampal regions, resulting in a global memory impairment. An MRI scan of J.C.'s left hemisphere can be seen in Figure 8. This shows the main area of brain injury plus some additional damage caused by surgery and by a metal clip inserted to prevent further haemorrhaging. Note that the damaged area is far less extensive than that seen in C.W.'s

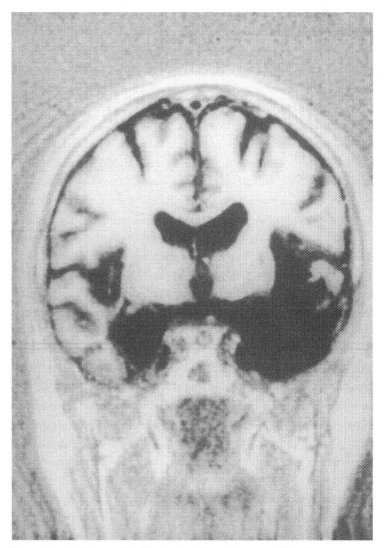

Figure 6 An MRI scan of C.W. showing bilateral temporal lobe lesions.

scan. Furthermore, J.C. had both frontal lobes intact, making it easier for him to compensate for his memory problems, since the frontal lobes are important for planning and organization.

Following hospitalization and rehabilitation, J.C. developed a sophisticated system of memory aids and strategies. In particular, he uses a dic-

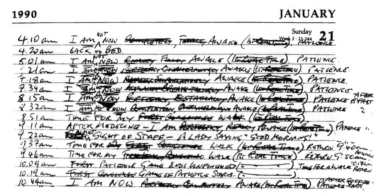

Figure 7 Part of one page of C.W.'s diary.

taphone, a Filofax and a watch with an alarm. He is very organized and tries to ensure that everything he needs to remember is recorded in at least two places. J.C. has been so successful that, despite a severe amnesia, he is able to live alone and earn his own living.

There are no doubt several reasons why J.C. has learned to compensate successfully while C.W. has not. J.C. is younger, the causes of memory impairment are different and the two individuals have different personalities. The most important reason, however, is the extent of damage caused to the brain: the profound destruction of brain tissue in C.W.'s case has produced not only a very severe memory disorder but also widespread additional problems that make it impossible for him to learn any compensatory strategies. J.C., in contrast, has a much purer amnesic syndrome; this cannot be cured but it can, to some extent, be managed.

Helping people with memory problems

There are a number of ways in which memory-impaired people can be helped to by-pass, manage or compensate for memory failures. As indicated earlier, one is simply to try to prevent difficult situations from arising in the first place, by restructuring or rearranging the environment to enable people to cope with limited memory functioning. Labelling the doors of each room in the house will, for example, direct people to those rooms without the need to remember. A similar procedure can be used

Barbara A. Wilson

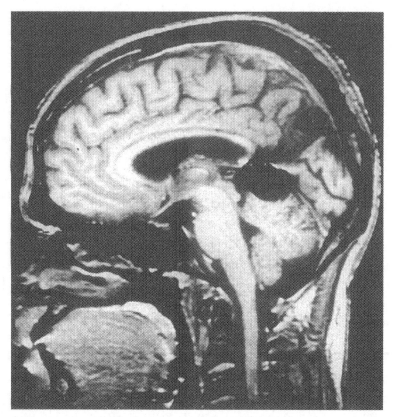

Figure 8 An MRI scan of J.C. showing hippocampal damage together with
some artefactual damage caused by a metal clip and some slight damage
caused by the surgery.

with cupboards and drawers. Indeed, some centres are already designing houses in which technological devices such as computers and video
links are employed to monitor and control the living environments of
people with dementia. These are known as 'smart' houses because they
contain a series of alarms and reminders linked to cookers, shower units,
telephones and other essential instruments for daily living. Individuals
are automatically reminded about necessary activities such as taking
medication, controlling temperature, switching off electricity, monitor-
ing visitors, closing doors and so on. For people with severe intellectual
impairments or with many cognitive problems besides memory impair-

ment (such as C.W.), environmental adaptations are probably just about the only effective form of help that we can offer.

People with less widespread problems, like J.C., can be taught to use external memory aids such as notebooks, Filofaxes, taperecorders and computers. Ironically, although external memory aids are often used to good effect by people without memory impairment, these devices are often very difficult for memory-impaired people to use efficiently because their operation *requires* memory. In other words, the very people who need external aids most have the greatest difficulty in using them. For example, they may forget to record information, forget to keep the aid with them, use an aid in a disorganized manner, fail to learn how to programme an electrical or computerized system, and/or be embarrassed at showing reliance upon aids. People like J.C. are exceptional: he is in many respects a 'star' subject who has taken several years to develop a systematic way of supplementing his memory performance.

One potentially exciting new aid, which avoids many of the difficulties outlined above, is 'Neuropage', a paging system designed specially for brain-injured people with memory problems. Designed by Larry Treadgold, the engineer father of a head-injured son, and Neil Hersh, a neuropsychologist, Neuropage is a simple and portable system usually worn on a belt. Reminders are programmed into a computer by a therapist. The computer is linked by telephone to a paging company, and once the messages are entered into the computer, no further human interface is necessary. On the appropriate date and at the appropriate time the information is sent to the individual's pager, which then beeps or vibrates to inform the wearer that a message is being transmitted. The pager itself is controlled by a single button, thus avoiding the need for alternative responses requiring memory of their functions. All the memory-impaired person has to do is press the button to receive the message. The same button can then be pressed to clear the screen and pressed once again to retrieve the message. This can be repeated as often as necessary, until a particular message is overridden by the arrival of a subsequent message.

For some people Neuropage has made a dramatic difference both to their own lives and to those of their families. As one man said when his wife was supplied with a Neuropage five years after her stroke, 'This is the best thing that has happened to our family for five years.' For other people Neuropage may not be the answer: one of our patients said to me, 'I don't like it; I feel as if Big Brother is watching me!'

Another area of technology that is likely to expand in the future is in the use of computers as prosthetic memory aids. Computers can be programmed to provide step-by-step instructions for coping with everyday life activities such as cleaning the house or cooking. Little knowledge of computer operations is required of the memory-impaired person, who follows instructions as relayed by a portable computer that can be wheeled round the house on a trolley. Studies employing these kinds of program report increased self-sufficiency and self-esteem as well as reduced stress.

Improving new learning

One of the biggest handicaps experienced by memory-impaired people is their difficulty in learning new information or skills. Memory is of course crucially involved in new learning, and impaired learning ability is frequently more of an impediment than failing to remember past conversations or events. A major goal of memory rehabilitation is to look for ways of helping memory-impaired people to learn more efficiently. Earlier on in this chapter, I mentioned the kind of learning that leaves no conscious awareness or recollection of when, where or how the learning occurred. Consider for a moment when and how you learned to read. The chances are that you will not be able to recall many or indeed any details of the process, although you can readily demonstrate the skill of reading. Examples like these are sometimes called implicit learning or memory, because the exercise of the knowledge requires no explicit memory for the events that established that knowledge.

Most amnesic people show normal or nearly normal learning on tasks involving implicit memory or learning. One such task is known as stem completion, in which a list of words is presented to the subject one at a time. Let us suppose there are sixteen words, with the first of them being *BEACON*, the second *CHEMICAL*, and so on. After seeing all sixteen words, an amnesic subject would be able to produce very few of them, because this is an episodic or explicit memory task where conscious recollection is required, and of course amnesic people have very poor episodic recall. However, if, after the person had seen all sixteen words, you presented the first three letters of each word, for example, *BEA* and *CHE*, and asked the amnesic subject to tell you the first word that comes to mind, the response is very likely to be '*BEACON*' or '*CHEMICAL*', despite the fact that these particular words are not the only nor indeed even the most probable completions of the stem (cf. *BEAN, BEAST, BEAVER, BEAUTIFUL*, etc.). The amnesic person would probably have no explicit memory of having seen *BEACON*, but the brain has clearly registered the earlier experience in some implicit manner.

Such implicit learning can, on the other hand, be harmful if you work backwards. Suppose you were to give the amnesic subject the letter B first of all, and ask for a six-letter word starting with this letter. The subject might say '*BUTTER*'. You reply in the negative and offer the first two letters, *BE*, and the subject replies '*BETTER*'. Again you say 'no' and offer *BEA*, to which the subject says, '*BEASTS*'. This process continues until the subject gets the right word, *BEACON*. Following this procedure, and after a short delay, if you were to present the first three letters and ask for the first word that springs to mind, the amnesic subject would be very likely to give one of the incorrect responses, such as '*BEASTS*', because there is implicit memory for that earlier (as it happens, incorrect) response as well as for the correct word. Thus, starting with the whole word and then presenting the stem leads to normal or nearly normal learning in memory-impaired people, while starting with a stem and building up the whole word results in impaired learning.

The conclusion that follows from this demonstration is that, because amnesic people cannot remember or distinguish their errors from correct responses and cannot therefore benefit from trial-and-error learning, it is better to prevent them from making errors when they are trying to learn new information. The very act of making an error is likely to reinforce it, in the absence of the kind of episodic memory that enables most of us to recall that the response X was right but Y was wrong.

Alan Baddeley, Jonathan Evans, Agnes Shiel and I have been conducting research for several years that is based on the question 'Do amnesic subjects learn better when prevented from making mistakes during learning?' In our first experiment, we used a stem completion task under errorless and errorful conditions with sixteen densely amnesic subjects, sixteen young control subjects, and sixteen elderly control subjects. In the errorless condition, subjects were prevented from guessing; in the errorful condition, they were required to guess and so could not avoid making mistakes. Every single one of the amnesic subjects learned more when prevented from making mistakes. There was a similar trend for better performance in the errorless condition by the control groups, too, though the advantage was not nearly so striking as for the amnesic group. We then carried out further studies employing the errorless learning principle for teaching practical, everyday tasks to amnesic people; and again, in almost every case, errorless learning proved to be superior to trial-and-error learning.

A final case study

I shall conclude by mentioning Jack, who sustained memory problems following a suicide attempt. Given the sad fact that all too few brain injured people receive rehabilitation in Britain, I think it is likely that the general reader will be unfamiliar with what actually goes on in memory rehabilitation, so I shall describe Jack's rehabilitation programme in a little detail.

Despite the fact that Jack's programme was relatively low cost, and did not involve sophisticated or expensive equipment, it resulted in considerable gains for Jack and his family. Jack and his parents were referred to me several months after the suicide attempt. Following an initial interview, in which members of the family were able to voice their distress, hopes and expectations, Jack was offered an assessment prior to rehabilitation. The assessment involved both neuropsychological testing, in which standardized tests were used to pinpoint his cognitive strengths and weaknesses, and a functional assessment to identify the nature of the everyday problems and the family's primary concerns.

Jack then received twelve weeks of individual therapy and several months of group therapy. The individual sessions took place for two hours each week. The family was given information and counselling about the nature of memory problems, and what to expect. In addition, Jack received training in the use of strategies to circumvent his problems. He also attended a weekly two-hour memory group attended by other memory-impaired people.

The group sessions included (a) discussions about problems and ways of overcoming them; (b) role playing in problematic situations; (c) practising the use of strategies and aids; and (d) playing memory games to encourage the use of personal strategies.

At the end of the twelve weeks of individual therapy, work trials were arranged for Jack at the Medical Illustrations Department of a nearby hospital. Jack did well there and was able to obtain a place at college to pursue a course in media studies. Although still amnesic, Jack has come a long way from the early days when his parents were given an extremely gloomy picture of his future. To end this chapter I would like to quote some of Jack's own words:

> My memory limitations are not so much a problem any more. I don't mourn the loss of my memory as I can't remember what it used to be like. The condition has helped me to evolve, I think, into a different type of person . . . I have become more cerebral, my thinking more esoteric, and I am very comfortable with this. I think it suits my character.

FURTHER READING

Baddeley, A. D., *Human Memory: Theory and Practice*, London: Lawrence Erlbaum Associates, 1990.

Baddeley, A. D., Wilson, B. A. and Watts, F. N. (eds.), *Handbook of Memory Disorders*, Chichester: Wiley, 1995.

Campbell, R. and Conway, M. (eds.), *Broken Memories*, Oxford: Blackwell, 1995.

Wilson, B. A., *Rehabilitation of Memory*, New York: Guilford Press, 1987.

Wilson, B. A. and Moffat, N. (eds.), *Clinical Management of Memory Problems, second edition*, London: Chapman & Hall, 1992.

7   How Brains Make Memories

STEVEN P. R. ROSE

The biological importance of memory

It is not brains that make memories; it is people, who use their brains to do so. And animals, non-human animals, also make memories, and can learn and change their behaviour as a result of experience. Even some animals without much in the way of brains at all, just rather basic nervous systems, can do it. What this points to is the tremendous importance that the capacity to learn and remember has for the survival of animals. Plants do not need nervous systems, because all they have to do is to stand around with their arms – or branches – spread wide so that their leaves can catch the sun and photosynthesize. But animals which live on plants, and even more so animals which live on other animals, have to use their wits to find and capture their prey, and to avoid being eaten in their turn at least long enough to be able to reproduce. Such ways of making a living in the world demand the development of sensitive sense organs, and the capacity to register and interpret the data provided by those sense organs, to compare it with past experience and, even more, with the outcomes of that past experience. And this is what learning and memory are all about. It is not the only route to evolutionary success. After all, bacteria do pretty well without either brains or nervous systems, or even much by way of memory – though there have been some disputed claims that they can learn from experience. But once evolutionary strategies based on brains appeared, the pressure to get bigger and presumably smarter brains, with more capacity to learn and remember, must have been substantial.

I shall begin with human memory, but I will be arguing quite strongly that we can learn a great deal about the brain mechanism of human memory, and even about strategies for repairing damaged memory, on the basis of studies of non-human animals. To me, memory is the feature that defines every single one of us as an individual. We can contemplate losing a limb or a sense, or even having a heart or kidney transplant, and still retain a conviction, albeit modified, of our own personhood. Imagine losing memory – or having a memory transplant *à la Manchurian Candidate* – and the difference is immediate and apparent. We *are* our memories. In old age we can retain memories of our own childhood eighty or ninety years previously, even though every molecule in our bodies, and every cell except our neurons, has been recycled many thousands, perhaps millions, of times. In many ways the problem of memory is similar to that of the problem of form – how is it that biological form is retained despite this constant demolition and rebuilding of our body substance? And of course this is why the diseases of memory, such as Alzheimer's, of which I will have something to say later, are so devastating.

Neuroscience and memory

It is in part for these reasons that, as a neuroscientist, I have been fascinated by memory for most of my working life. I am constantly intrigued by the way in which a feature of our own existence which is so intensely personal, individual and subjective can none the less be explored by the 'objective' methodology of science. The sense of confidence, even arrogance, amongst the industrial world's neuroscientific community is tangible. Around 25 000 researchers meet each year at the annual jamboree of the American Society for Neuroscience; our once recondite research field has become big business for pharmaceutical companies and putative genetic engineers. Bandwaggoners, from molecular biologists and quantum physicists to artificial intelligence (AI) modellers, are offering to tackle the Big Question, that of Consciousness. My aims are more modest. I have argued before and will repeat here that the study of learning and memory may be the key to deciphering the translation rules which lie between the languages of brain and mind, the Rosetta Stone of the neurosciences. There is a methodological principle operating here. In science it is always easier to study change than stasis. When a person or an animal learns, their behaviour changes in specific ways, and one can then ask what changes in the brain to match, to correlate with, the change in behaviour. This approach, which I will spend part of this chapter attempting to illustrate and justify, has lain at the heart of my own experimental programme over the past getting on thirty years.

The limits to memory

But let us begin at the beginning, with the study of the limits of human memory itself. Fascination with the strengths and weaknesses of human memory go back to ancient times. Frances Yates, in her book *The Art of Memory*, described the mythic origins of the Greek and Roman techniques for cultivating memory. The poet Simonides is commissioned to sing the praises of his host at a Greek banquet, but makes the mistake of praising the twin gods Castor and Pollux at the same time. When the time comes for him to be paid, his host offers him only half, suggesting he get the rest from the gods. Just then, Simonides is summoned from the banqueting hall by two young men who turn out (surprise, surprise) to be Castor and Pollux themselves. In his absence the hall collapses, crushing host and guests beyond recognition. The grieving relatives ask Simonides to help to identify the corpses, and he is able to do so by recalling the order in which the feasters were sitting round the table. This became the basis on which Greeks and Romans cultivated what they called artificial memory. To recall (in the days before notepaper and autocues) a speech you were to give in the Senate, you decided on the points you wished to make and then 'placed' them close to objects in a familiar environment so that you could recover them in sequence as you walked, in your imagination, through that environment.

This technique, surprising though it may seem, works rather well. Yates traced it through the 'memory theatres' (Figure 1) which culminated in the hermeneutic tradition of Giordano Bruno; but if even today you answer those advertisements in the Sunday newspapers which offer you methods for improving your memory, you will find yourself offered a similar method. The current holder of the record for remembering pi to 4000 places – or whatever the figure is – told me he used a similar technique, but believed he had invented it himself. And the Russian neuropsychologist Alexander Luria, in his extraordinary little book *The Mind of the Mnemonist*, describes how one patient, a man made desperate by an incapacity to forget, remembered nonsense formulae Luria had given

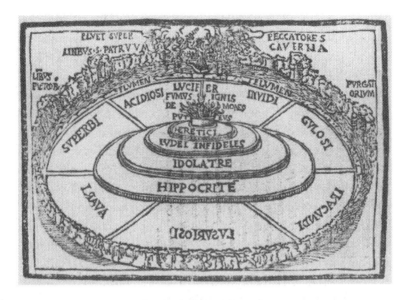

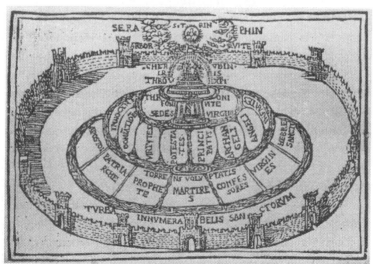

Figure 1 A memory theatre. *Above*: Hell as artificial memory.
Below: Paradise as artificial memory.

Steven P. R. Rose

him over a period of practically two decades. Indeed Luria's case illustrates another point about memory – forgetting is functional. To remember all the data which pass through one's senses every day would be impossible; your memory would become, in the words of one of Jorge Luis Borges' memorable (*sic*) characters, 'a garbage disposal.'

Graeco-Roman experiments with artificial memory also reveal another feature of the way in which the study of memory has proceeded over the centuries. Perhaps more than other sciences, biology has always worked with metaphor, and throughout history the brain and its properties have been analogized to whatever have been the best technological artefacts available to the culture. For the Greeks and Romans, memory was inscribed on wax tablets. Later, in the Renaissance, clockwork or hydraulic metaphors emerged. By the eighteenth and nineteenth centuries, the brain was a telegraph or telephone exchange. Today of course it is a computer, as Terry Sejnowski discusses in Chapter 8, but I am going to suggest another way of modelling memory.

Suppose I flashed a set of seven numbers on to a screen and asked you to repeat them back. You would probably do so without fail. How about eight numbers? It gets harder. Nine, and most people are lost. What? Unable to remember a sequence of eight digits? A simple calculation shows that to remember this requires a mere 41.86 bits of information. A basic pocket calculator has a 1000 bit memory, and the floppy disc of the Macintosh on which I am typing this more than ten million. So computers win hands down. Yet I can easily remember and read off to you a forty digit long string of numbers, and I am not Luria's patient or winning a place in the *Guinness Book of Records*. How do I, and others, do it? Simple: the numbers are sequences which I know as telephone numbers or birthdates.

I don't view computer memory as a metaphor for human memory. As I see it, computers deal with dead, static *information*, locked into files, that can be pulled out, inspected and replaced. Humans deal not with information but with *meaning*. How many bits of information are there in my memory of my fourth birthday party, the colour of my grandson's hair or even what I had for breakfast this morning? Incomputable. Yet the AI people persist in confusing them, as when my fellow memory researcher at Oxford, Edmund Rolls, calculates that the primate hippo-

138

campus can store precisely (I like the precision) 36 500 distinct memories. I do not have to go all the way with the mathematician Roger Penrose into quantum speculation to dismiss the presumed relationship between computer memory and human memory as not much more informative than a wax tablet or hydraulic metaphor. Which is incidentally why I did not mind whether the computer Deep Blue did or did not beat Gary Kasparov at chess. Chess is a game of 'pure' cognitive analysis. Human memory, human action, demands not just cognition but affect too – which is why it may be fun to play chess against a computer, but pretty boring to play poker. I would like to replace the famous Turing test for human-like computers with a poker test instead. I made the personal transition between chess and poker as an undergraduate in the late fifties, and I have never looked back.

The taxonomy of memory

This business of remembering numbers illustrates some other features of human memory too, which will turn out to be important once we begin to look at mechanisms. People forget a seven figure number quickly unless it is an important one such as a telephone number. There is a temporal gradient in memory – some things are remembered only for a short time, others for much longer. There is a great deal of evidence that the transition from short- to long-term memory is a crucial step, both in terms of function, molecular and cellular mechanisms, and what might go wrong in disease states. *Short-term* versus *long-term* memory is the first of the dichotomies in memory studies.

This is another. Seven or eight numbers may be all you are able to recall, but suppose one tests your memory in a different way. Consider an experiment done by Lionel Standing, a Canadian psychologist, in the 1970s. He showed groups of subjects sequences of photographs, one after another, each for a few seconds. A week later, he called them in again and now showed them pairs of photographs – one they had seen before, and the other novel – and asked them to identify the one they had previously seen. His subjects could recognize up to 10 000 photographs with 90% accuracy. So *recognition* memory seems quite different from *recall* memory – and if that surprises you, think of the difficulty we all have in describing even a well-loved friend's face compared with the ease with which we recognize the person when we see them. We remember abstract information – such as numbers – differently from how we remember complex scenes or patterns.

Inferring function from dysfunction

The taxonomy of human memory (Figure 2) owes much to the study of accidents of nature, brain lesions caused by stroke or damage or degenerative diseases such as Alzheimer's. The generally accepted taxonomy is due to the San Diego neuropsychologist Larry Squire, who distinguishes first between *declarative* and *procedural* memory – knowing *that* and knowing *how*. Knowing that an object with a saddle, two wheels and handlebars is called a bicycle is declarative; knowing how to get on it and ride it is procedural. These two types of memory seem to rely on different cellular, brain and bodily mechanisms, for the first is much more readily lost than the second (you can ride a bike even after a twenty year gap with almost no need for rehearsal). Declarative memory subdivides into *semantic* (knowing what doctors do) and *episodic* (knowing that I went to the doctor's last Tuesday). This latter is the most vulnerable in conditions like Alzheimer's.

Accidents of nature provided until relatively recently virtually the only way to study the processes of human memory. In a recent book, *Rewriting the Soul*, the philosopher and statistician Ian Hacking dates the birth of the sciences of memory to the 1870s in France, with the recognition that there were specific diseases of memory. The study of recovery of memory after blows to the head – concussion – led to the recognition of the relationship between short- and long-term memory. It appears that, after such a blow, memory for events up to a few minutes prior to the accident can be recovered but these last moments appear to be irretrievably lost – hence the argument that memory must be 'stored' in the brain in a more vulnerable form. Once past this critical time memory seems relatively invulnerable to such accidents.

Perhaps the most famous and prolific exploitation of an iatrogenic accident in the history of medicine relates to a man known in the literature only by his initials, H.M. Operated on in the 1950s in Montreal for epilepsy, H.M.'s hippocampus (Figure 3) and portions of his temporal lobe were removed. The result was that H.M., whilst retaining his mem-

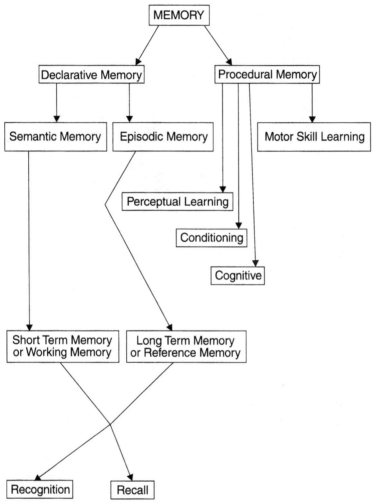

Figure 2 A taxonomy of memory.

ories up until the time of the operation, has forever lost the capacity to transfer new memories from short- to long-term store. He remembers events only for a few minutes, before they fade from history. Every day begins afresh for him, back in a mental timewarp in which he is still a young man with the interests, attitudes and knowledge that he had before the operation was carried out.

From this neuropsychologists concluded that the deep, neuron-rich

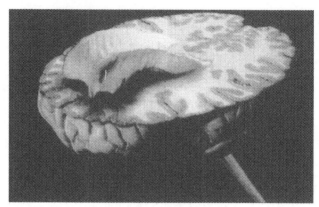

Figure 3 The hippocampus in the human brain.

region of the cerebrum, called the hippocampus for its fancied resemblance to a seahorse, was a structure essential for memory, or at least for the registration of short-term memory and its subsequent transfer to long-term store. The hippocampus has remained a controversial focus of physiological attention ever since. But a word of caution is required. Inferring function from dysfunction is notoriously difficult, even for inanimate objects. As Richard Gregory pointed out many years ago, if you remove a transistor from a radio and it emits a howl rather than a symphony, you cannot conclude that the function of the transistor is to be a howl-suppressor. You are studying not the absent part but the system in the absence of the part. And the dynamic plasticity of living systems, the multiply redundant structures and functions of the brain, make what is true even for transistor radios dramatically the case for brains.

Normal memory in humans and animals

So we need to study functional as well as dysfunctional memory. The classic psychological studies of memory begin with Hermann Ebbinghaus, a German psychologist of the nineteenth century, who set himself to learn, and later to recall, lists of nonsense syllables. Such studies confirmed that the loss of memory occurs mainly in the first hour or so, after which what is retained is relatively stable. The study of normal human memory has remained a fertile field ever since, but in a strange way fails almost entirely to connect to the neurobiological literature. The most authoritative recent book on human memory, called just that, is by Alan Baddeley. It contains virtually no references in common with the standard neurobiological books on memory. It concentrates on memory in everyday life, on working memory, on forgetting, on strategies for remembering, on so-called 'flashbulb' memories and on autobiographical memory. None of these translates readily into the laboratory languages used in neurobiological circles. Despite all the claims for the convergence of neuroscience and psychology over the past decades we clearly still have a long way to go.

Perhaps that is because almost at the same time as Ebbinghaus in Germany was founding the study of normal human memory, Ivan Pavlov in Russia was beginning to explore the physiology of animal memory. Not merely could his dogs learn – any shepherd could have told him that – but they could do so under standard, quantifiable laboratory conditions. The study of classical conditioning began with the famous dog salivating to the sound of the bell which heralded the arrival of its food. Researchers began to ask questions such as how long could the experimenter extend the interval between bell and food and still achieve the 'association', could one extinguish the association by sounding the bell and not bringing the food, and so forth. And a few years later into the present century, along came J. B. Watson and B. F. Skinner and gave us behaviourism and operant conditioning, in which the animal learns to do something like press a lever or move to a particular site in its cage to obtain a 'reward' of food or water, or to avoid 'punishment' such as an electric shock. Skinner called these 'contingencies of reinforcement' and offered to turn psychology into an exact science, a physics of behaviour, in which the brain was a black box whose internal content was irrelevant to the output of the animal in response to these varied contingencies. Pavlov and Skinner ruled unchallenged until the birth of modern neuroscience and cognitive psychology in the 1960s; today we are interested in these previous approaches mainly as historical blips; both approaches to understanding memory have proved scientifically sterile.

It is literally only within the last decade that new methods to explore the functioning human brain have emerged, with the development of powerful and relatively non-invasive imaging techniques, such as PET

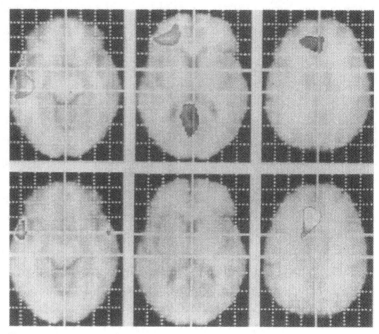

Figure 4 PET images of memory. Regions of the brain showing increased blood flow during learning and recall of word lists as demonstrated by positron emission tomography.

(positron emission tomography) and MEG (magnetoencephalography). Neither is perfect: PET (Figure 4) gives good localization but not in real time, and because it involves radioactivity cannot be used repeatedly on the same subject; MEG (Figure 5), in its infancy as yet, gives superb time-resolution but poor spatial location. Both are, however, beginning to reveal insights into the dynamics of learning and memory. For instance, when a person is asked to learn and recall a word list, the initial learning processes activate regions of the left frontal cortex – that is, these regions show enhanced blood flow and glucose and oxygen utilization, taken as surrogate measures for neural activity. By contrast, when the subject is asked a brief time later to recall the same list, regions of the right hemisphere become engaged. The problem with all these measures at the moment is that, dramatic and beautiful as their resulting pictures may be, we do not know yet what they mean in terms of the brain structures, chemistry or the detailed functions involved.

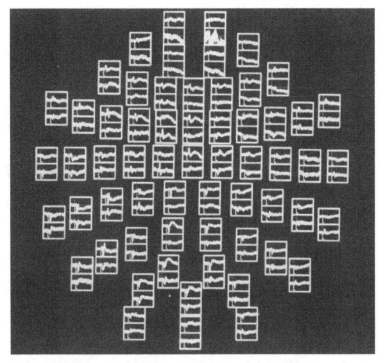

Figure 5 MEG images of memory. Magnetoencephalographic images drawn from the brain of the author whilst learning a complex recognition task. Each trace shows a few milliseconds of record from a separate channel in the machine; the channels are distributed across the skull.

Hebbian synapses

To go beyond this means to enter the brain itself, to explore the world of its nerve cells and the connections between them, the synapses. All of us working in this terrain owe our theoretical perspective to the Canadian psychologist Donald Hebb, who in 1949 published his influential book *The Organization of Behavior*. Hebb was impressed by the durability of memory once past its short-term phase. Such stability, resistant to most brain insults – electroshock, insulin coma and the like – must depend on permanent changes in the circuitry of the brain, on its synapses (Figure 6). He argued that the making of memory depended on the physical growth or realignment of synapses, thus creating novel pathways. Given that there are perhaps a hundred billion nerve cells (neurons) in the

Steven P. R. Rose

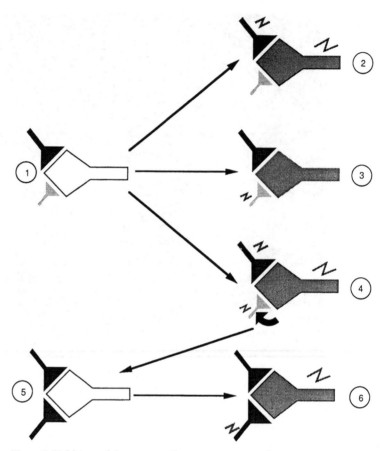

Figure 6 Hebb's model synapses. Shown are synapses from two neurons onto a third. Initially only the larger synapse is strong enough to cause the third neuron to respond by firing. But if both the larger and smaller neurons fire simultaneously, feedback from the third neuron to the smaller synapse results in it increasing in size and strength so that now it is also independently able to cause the third neuron to fire.

human brain, and something like 10^{14} synapses between them, that is a huge number of possible circuits.

Hebb's ideas still dominate the field, though with modifications. Few now believe that memory occurs as a result of changes at a single synapse, although this case was put forward quite strongly for forms of learning in the sea mollusc *Aplysia*. (*Aplysia* is a somewhat overgrown slug with a limited but measurable behavioural repertoire, which became a favoured neurobiological model for memory studies during the 1980s, but has now somewhat lost its earlier prominence.) Instead, those memory theorists influenced by AI talk about the properties of ensembles of neurons, linked by networks of synaptic connections, in which memories are stored in the form of patterns of distributed alterations in synaptic strengths, or neural nets. This is the approach taken by Sejnowski in Chapter 8, and the basis for Rolls' claim that 36 500 memories can be stored in the hippocampus. For 'memories' read discrete patterns of connections, granted certain assumptions about the numbers of neurons and of synapses available. But I believe such models are still a long way from biological reality. When computer people speak of neural nets, they do not mean the neurons and synapses I can see down my microscope, but models made in their computer systems.

During the sixties and seventies, several laboratories were able to show that rearing animals – rats or cats for instance – in specific environments would alter the growth and pattern of neural connections. For instance, when rats were reared in the dark and subsequently exposed to light, this change in environment resulted in a dramatic increase in synapses visible even in the light microscope (see also Figure 7). Doing the same with cats changed the physiological properties of cells in their visual brain regions. Similarly, a famous series of experiments in which rats were raised in so-called 'enriched' versus 'impoverished' environments showed that quite dramatic changes in brain structure and chemistry were possible as a result of experience. But such general examples of plasticity were still a long way from showing that memory formation itself resulted in specific cellular changes. And a number of false starts and irreproducible experiments – such as those claiming that it was possible to isolate specific 'memory molecules' (proteins or RNAs) – achieved popular notoriety whilst bringing the research field into disarray and disrepute.

The chick as a model system

What the study of the cellular mechanisms of memory required was a reliable animal model of learning, to which the sophisticated modern techniques of molecular biology, neurophysiology and image analysis

Steven P. R. Rose

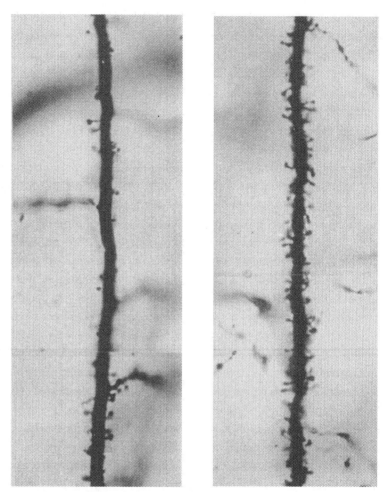

Figure 7 Growth of dendritic spines. Apical dendrites from the visual cortex of littermate rats, the first (*left*) reared in the dark, the second (*right*) after a few days' exposure to light.

could be applied. Several now exist, but I shall discuss only one, that with which I have been working for many years now. My introduction to the day-old chick as a potential subject for learning and memory experiments came through a collaboration which lasted over a decade with two distinguished colleagues from Cambridge University, Pat Bateson and Gabriel Horn. Together, we explored the biochemical events that

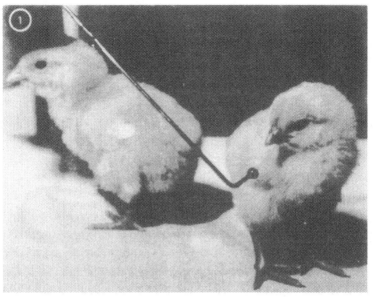

Figure 8 Pecking chicks.

occurred in the young chick's brain during the process of imprinting; that is, as it learned to distinguish an object which it would regard thereafter as mother. Whilst Pat and Gabriel and their co-workers have continued to work very successfully with imprinting in the years that followed, from around 1980 I decided that for my purposes a simpler form of learning made a more appropriate system.

Young chicks explore their environment energetically. They have much to learn, and need to do so fast. This means that by contrast with the complex studies of sophisticated forms of learning of mazes or lever-pressing tasks in adult animals, in studies of this kind of early learning one is working experimentally with the grain of the biology, allowing the animal to learn quasi-naturally. If offered a small bright object such as a bead, chicks will peck vigorously at it within a very few seconds (Figure 8). If the bead is made to taste bad by dipping it in a bitter liquid, the chicks will peck once, shake their heads vigorously, wipe their beaks on the floor of the cage, and avoid beads of a similar size and colour for several days subsequently. This is called one-trial passive avoidance learning. In contrast, chicks which have pecked a dry or water-coated bead will continue to peck at it when offered it subsequently. The training is both rapid and reliable, an essential prerequisite for the analyses that follow. One can then ask the question: what happens in the brain of the chick that has experienced the bad-tasting bead, and which can be taken to constitute the neural representation of that experience – its memory? Since 1980, I and my colleagues at the Open University have brought much of the armamentarium of modern biology to bear on this question, and I want now to summarize, without going into excessive biochemical detail, the substance of our findings.

Correlation and intervention

There are two broad approaches. The first is *correlative*; one can train the chicks on the unpleasant-tasting bead and observe differences from the control chicks. Of course, these questions are not asked in a spirit of total biochemical naïveté. We have reasonably clear ideas about the sort of molecular processes that might change. But such a correlative approach is open to the charge that any change we observe might not be a consequence of the laying down of some type of memory trace, but might instead be a consequence of the sensation of the bitter taste of the bead, or the visual or motor activity associated with the experience of pecking. A good bit of our work has been devoted to designing control conditions to eliminate the effects of these factors – a clear-cut example of a reductionist methodology at work.

The second approach is *interventive*. If a particular biochemical process – for instance the synthesis of certain types of protein – is *necessary* for memory formation, then preventing that process, for instance by the use of a drug or metabolic inhibitor, should prevent memorizing processes from occurring. That is, if treated in this manner the experimental chick should show amnesia and, instead of avoiding the bead if it is subsequently presented, should peck it instead. Once again, controls are required to demonstrate that the effect of the drug is directly on the formation of the memory trace, and not on some other aspect of its expression – for instance, making the chicks excitable and prone to peck at anything. The goal of this approach might be summarized as the attempt to discover the necessary, sufficient and exclusive brain corre-

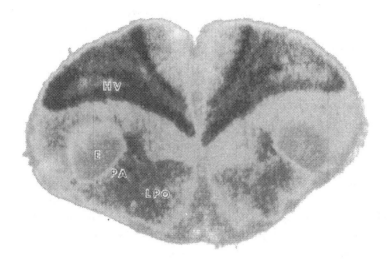

Figure 9 Memory storage sites in the chick brain. Cellular events following training are initiated in the left IMHV and later also appear in the right IMHV and both left and right LPO.

lates of memory formation – a task I have discussed at some length in my book *The Making of Memory*.

Sites of change

It was first necessary to show that something changed in the brain as a result of training. The least loaded physiological assumption to make is that any neuronal restructuring will demand an increased use of energy, and hence glucose, the brain's major energy source. One can use a method analogous to the PET studies in humans, in which a radioactively labelled metabolic analogue of glucose is introduced into the blood supply and its accumulation in different brain regions measured. This technique showed us that two regions of the chick brain in particular 'lit up' when the bird was trained on the passive avoidance task. These regions have the usual cumbersome dog-Latin neuroanatomical names; here we need only refer to them by their initials – IMHV and LPO (Figure 9; the IMHV is the medial region of the part labelled HV in the

Steven P. R. Rose

slide). Because the avian brain is lateralized, with different functions localized to left and right hemispheres, it was not very surprising to us that the region of change *par excellence* was in the left hemisphere – the left IMHV. This now gave us a site to explore in more detail.

Sprouting synapses

Do synapses actually change in this region? My long-term colleague in these projects, Mike Stewart, studied this question using both light and electron microscopy. The short answer is 'yes': twenty-four hours after training chicks on the task, there is a dramatic increase in the numbers of specific types of synaptic connection in both IMHV and LPO, though this overall change masks some quite complex time-dependent dynamics. What seems to happen is an initial sprouting of large numbers of new connections. Indeed, some are formed as rapidly as half an hour after the training episode and then, as time goes on, become pruned away, leaving what we may presume to be only the minimum necessary for recall or recognition of the bitter bead. Furthermore, the new connections are not made stably in a single small ensemble or brain region, but are widely distributed and appear in sequence in different regions at different times.

If the new synapses are physiologically 'meaningful' we might also expect that their electrical properties would change as a consequence of training. This too proves to be the case. If chicks are trained on the passive avoidance task, and later anaesthetized and a recording electrode – essentially a fine wire – placed into the IMHV or LPO, we pick up characteristic patterns of novel neural electrical activity, bursts of high frequency firing by the neurons. These burst patterns also shift in space and time, beginning in the left IMHV and later shifting to the right IMHV and LPO. In particular – and this is important for what follows – there is a specific time interval, about six to eight hours after the training, when the bursts are particularly strong and during which their focus shifts between brain regions.

The biochemical cascade

I am a biochemist by training, and the topics that have primarily concerned me have been to map the cellular and molecular events that correspond to these changes detectable by the techniques of neuro-anatomy and physiology. I am only too well aware, however, that biochemical processes which are music to my ear sound cacophonous or even produce functional deafness in others who do not share my enthusiasms, so I am going to cut a very long story very short indeed.

In the minutes to hours following the training of chicks on our bead-pecking task, a veritable cascade of biochemical reactions occurs in the IMHV and, subsequently, in the LPO. Within minutes of pecking, the neurons of the IMHV release a surge of a particular neurotransmitter – the chemical which carries signals between the pre- and postsynaptic neurons. The transmitter concerned is an amino acid, glutamate, and as it crosses to the postsynaptic side there is a matching increase in the availability of the receptor proteins embedded in the synaptic membrane, which receive the transmitter and transduce its effect into a signal of excitation for the postsynaptic cell. The arrival of this message stimulates a flurry of biochemical activity in the postsynaptic neuron, which I will come to in a moment.

At the same time, lasting changes in the connectivity between the two cells require changes in the presynaptic side as well. There had for some time been speculation that the signal to initiate these presynaptic changes must be some type of so-called 'retrograde messenger' molecule which was released postsynaptically and flowed back to the presynaptic side. And in accord with these speculations, about three years ago we were able to show that the putative retrograde messenger, the simple gaseous molecule nitric oxide, did indeed play a role in establishing memory. If the production of the nitric oxide is blocked by injecting a simple metabolic inhibitor, the chick will learn and remember the bead-pecking task for a brief time – between fifteen and thirty minutes – and then, almost abruptly, forget it. The techniques of correlation and intervention have enabled us to dissect out a complex, time-dependent cascade of processes (Figure 10) occurring in the first hour following training on the task. By choosing the appropriate drugs or inhibitors, we can in essence stop the biochemical cascade at any point, and by doing so block any subsequent memory for the task.

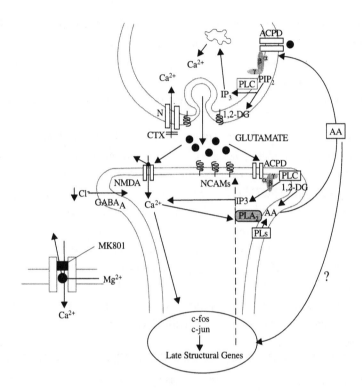

Figure 10 The biochemical cascade of memory formation, giving some of the sequence of synaptic events in the minutes to hours following training. The upper part shows the presynaptic reactions leading to calcium-controlled release of the chemical neurotransmitter (in this case glutamate, indicated by black dots), initiated by a change in the electrical state of the neuron. The lower part shows the action of the neurotransmitter on the postsynaptic membrane, giving rise, through a sequence of receptors, second messengers and lipids, to gene activation and protein synthesis.

Cell adhesion molecules

Lasting changes in synaptic strengths, however, cannot rely on transient changes in transmitter flux. At both pre- and postsynaptic sides, the signals initiated by the transmitters and retrograde messengers culminate in the activation of genes and the synthesis of new proteins. Indeed, one of the best-established findings in the whole literature of the biochemistry of memory formation is that preventing protein synthesis blocks long-term memory, and this is certainly true in the chick too. Gene expression itself goes through at least two phases. The first, which is initiated about thirty to forty-five minutes after the training experience, involves the activation of a class of genes known as transcription factors, or immediate early genes, of immense interest to those elucidating the complexities of molecular biological housekeeping, but which need not detain us further here.

What is important, both theoretically and, as I hope to convince you, practically, is the class of proteins whose synthesis follows that of the early genes, for it is these proteins which are responsible for both maintaining and changing the configuration of the synapses. Pre- and post-synaptic sides of the junction between neurons are held together by proteins known as cell adhesion molecules, whose importance was first identified by the Nobel prizewinning immunologist Gerald Edelman. These molecules span the synaptic membrane, with a tail which lies in the interior of the cell and an external head end which projects out of the cell into the extracellular space. This head end consists, like all proteins, of a long sequence of amino acids, but unlike most other proteins, chains of sugar molecules are attached to the amino acids. Such complex amino acid/sugar macromolecules are collectively known as glycoproteins. The sugars confer a particular property on the glycoprotein – they make it sticky. This means that the extracellular portion of one of the cell adhesion molecules sticking out from the presynaptic side of the junction can glue itself to the matching sugar sequence of another cell adhesion molecule sticking out from the postsynaptic side, thus holding the two sides of the synapse in place.

What we have been able to show over the last three years is that, a few hours downstream of the initial training experience – the bead-peck – there is a surge of synthesis of cell adhesion molecules which are transported to the 'activated' synapses and inserted into them, thus, presumably, stabilizing them in their new configuration. Blocking the synthesis of these molecules, or preventing them sticking to one another by introducing into the brain specific antibodies or specially tailored small peptides which will interrupt the binding process, results in amnesia. The onset of this amnesia occurs at a period, about five to eight hours after training, which coincides exactly with the increased high frequency electrical bursting activity I referred to earlier and with the apparent shift in the 'memory sites' between IMHV and LPO.

Universal mechanisms?

Where do these findings lead us? First, the molecular processes that I have described are not unique to the chick. A major biological principle is one of biochemical parsimony. The cellular and biochemical processes that occur in species as diverse as *Aplysia*, chicks, rats and humans are very similar. The most expert neuroanatomist could not tell, from an electron micrograph of a neuron in the IMHV of the chick, whether it had come from a bird or a primate. There is every reason to believe that whilst the initiating events that trigger the biochemical cascade in the chick will differ from those in you or me when we learn a number sequence or are introduced to a new colleague, those that occur downstream, the molecular biological processes and involvement of cell adhesion molecules, are likely to be common across species and types of learning. Indeed, in collaboration with colleagues in Paris, our group has been able to show that antibodies we have made to cell adhesion molecules in the chick will also produce amnesia for quite different learning tasks in the rat, when given at exactly the same time (5.5 hours) post-training.

Distributed memory systems

It is important also to emphasize that although I have talked about biochemical processes and molecules, the memory that we are discussing does not in any sense reside in these molecules. There are no unique memory molecules. The memory belongs to the organism, the bird or the human, as a whole. Within the brain, the molecular processes that we are observing change the organization of specific connection patterns in diffuse regions, and insofar as the memory has a residence at all, it lies in these patterns; it is a property of the system as a whole. Indeed, one can go further. Although the processes I have described give the appearance of supporting the idea that the memory circuits are in some sense hard-wired once the new patterns have been established, other experiments from our laboratory, which I do not have space to discuss here, make it clear that this is not the case. The memory 'store' is not fixed but dynamic. Furthermore, it does not reside in a discrete brain location. Indeed, 'it' is not a single entity at all. When I train a chick, I present a bead to the bird, and perceive it myself, as a single unified object; but it is clear that the bead is not represented this way in either my brain or that of the chick. Brains classify data presented to them. In the case of a bitter bead, classification may be by colour, shape or size, locale and context of presentation, and of course taste. When the chick first tastes the bitter bead it is not to know which of these features of the experience are important, so it will tend to classify all of them. But the sites of classification within the brain are not coterminous. We believe, for instance, that colour representation may reside in the IMHV, whilst shape and size may be represented in the LPO.

Research from many sources points to the modularity of the brain. For instance, the primate visual system consists of, on present counts, some fifty or more separate areas (beautifully described in Semir Zeki's book *A Vision of the Brain*). At one time it was believed, without much evidence, that all these areas were organized hierarchically, each reporting up to some 'homunculus' in the brain which viewed them all in what Dan Dennett has memorably characterized as a Cartesian theatre. However, it turns out that this is not how brains work. There are no 'grandmother neurons' which have the task of integrating inputs from many regions so as ultimately to decide, yes or no, this is or is not your grandmother. Instead, representations of grandmothers and bitter beads are distributed. The brain is not a Stalinist command economy, but a well-integrated anarchistic commune. A major current theoretical problem for neuroscience is to understand how these distributed properties can be organized in such a way as to provide coherence, so that we, and chicks too presumably, perceive the world in a unified manner. The partial representations of fragmentary characteristics of the bead or the grandmother must come together. This has become known as the 'binding problem'. I will not discuss here the efforts to solve this crucial problem, except to hint that the high frequency neuronal bursting that I have talked about, and related oscillations that others have observed in mammalian studies, may have something to do with it.

Steven P. R. Rose

Cell adhesion molecules, memory and Alzheimer's disease

The third implication of our work is of more than merely intellectual interest. So far, I have described the effect of training the chick by offering it a strong bitter bead to peck and taste. This memory, as I have said, persists for at least several days. However, if we train the chick on a much weaker stimulus, a less bitter taste such as that of quinine, the chick will remember and avoid the bead for only a relatively short time, and then apparently forget, so that it pecks rather than avoids a dry bead when tested. The cut-off point for this short lasting memory appears to be about eight hours – that is, the time when, with stronger learning, the cell adhesion molecules are made and inserted into the synaptic membranes (Figure 11). We have evidence that, in weak learning, the 'second wave' synthesis of the cell adhesion molecules does not occur. However, it is possible to stimulate the synthesis of these molecules by various drugs, in particular, as my colleague Carmen Sandi has shown, steroid hormones (for which there are receptors in the brain). And the dose of the steroid hormone which stimulates the synthesis of the cell adhesion molecules will also convert the weak learning into strong learning and enhance the retention of the memory.

Why does this matter? Earlier in this chapter I have referred to the tragic consequences of Alzheimer's disease, characterized by a progressive loss of memory. Much effort has been devoted to developing drugs which might alleviate this memory loss, but to date most of the attention has been devoted to agents which affect neurotransmitters. For some time I have believed that this may be the wrong place to look. Recent molecular biological studies have demonstrated that one of the key molecules whose metabolism is disrupted in Alzheimer's disease is a substance known as the amyloid precursor protein, APP. In the disease, parts of this molecule get chipped off and accumulate in the spaces between synapses, resulting in the characteristic plaques which can be seen in the brains of Alzheimer's sufferers. Now it turns out that the APP belongs to the class of cell adhesion molecules, and its extracellular head with attached sugars presumably plays a part in the sticking together of synapses that I have described. I do not yet know how directly APP is involved in these processes, but, if it acts like those adhesion molecules

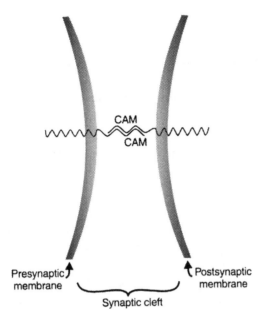

Figure 11 Cell adhesion molecule (CAM) binding pre- and postsynaptic membranes.

we have already studied, it becomes obvious why one of the early deficits in Alzheimer's disease is loss of the ability to store new long-term memories. And we may be able to extend our observations of the effects of hormones on both the synthesis of the molecules and memory retention by developing a class of pharmaceuticals which, acting via APP, may also be beneficial in the treatment, though not alas the cure, of Alzheimer's disease.

Real memory and neurobiologist's memory

I have tried to give you a flavour of research into the neurobiology of learning and memory, and to convey some of the reasons why I am enormously excited by what we and many other laboratories are currently discovering about the cellular and molecular events that occur when animals learn. Have I written anything at all about memory in the sense in which other contributors to this book, Richard Sennett or Jack Goody or Antonia Byatt, understand the term? To tell truth, I am far from sure. The distinguished psychologist Endell Tulving has pointed out, not unfairly, that when most neurobiologists talk of studying memory, in fact we are studying learning and memory formation. We know much less – almost nothing at all indeed – about what happens in the brain during recall of already learned experiences. It is glibly assumed that the patterns generated in the initial learning experience are reactivated, like retreading a familiar pathway, but we have no real evidence of the processes involved. Indeed it is difficult to know how one might study this in non-human animals. What will be feasible is much more detailed and sophisticated research using PET or MEG in humans who, after all, can be asked to recall in ways in which it is impossible to interrogate non-humans.

What is clear, however, is that brain memory is dynamic and not static, that the act of recall is not a passive one of pulling material out of some computer file, but an active process. Re-membering, as in particular feminist scholar Gayle Green has pointed out, is hard work. Re-membering is itself a biological process, and each time we re-member, we recreate our memories. Thus in a sense when we re-member we are recalling not the original event but our previous memory of it, itself transcribed into patterns of neural circuitry within our brains. That's one reason why, for a biologist, the debate about memory discussed by Juliet Mitchell in Chapter 5 is so interesting. I would argue as a biologist that there is no such thing as a false memory. There may be many memories which do not conform to historical, verifiable truth. But true or false, these memories have become inscribed in our brains, and cannot readily be altered.

I started this chapter by claiming that one reason for my preoccupation with memory is that it seems to stand at the juncture between the objective world of the laboratory and the subjective world of our lived experience. I confess that I am still a long way from unifying my understanding of what goes on in the synapses I study for a day-to-day research living with the cascading memories that I carry with me in my inexorable journey from cradle to grave. I am convinced that if I had a personal cerebroscope, wired up through life to a portable MEG machine or whatever, I could make this connection. On the other hand, however well I understand the circuitry and neural patterns, I am also pretty sure that the content of my memory will always be coded in the language of my mind, and not in the brain language to which it undoubtedly corresponds. Of that language, poets and novelists, not neuroscientists, must speak, for it is they who have the gift of tongues.

FURTHER READING

Baddeley, A., *Human Memory: Theory and Practice*, London: Lawrence Erlbaum Associates, 1990.

Dudai, Y., *The Neurobiology of Memory*, Oxford: Oxford University Press, 1989.

Hebb, D., *The Organization of Behavior*, New York: Wiley, 1949.

Luria, A. R., *The Mind of a Mnemonist*, London: Jonathan Cape Ltd, 1969.

Rose, S. P. R., *The Making of Memory*, London: Bantam Press, 1992.

Shaw, G. L., McGaugh, J. L. and Rose, S. P. R. (eds.), *Neurobiology of Learning and Memory* (reprint volume of papers), Singapore: World Scientific, 1990.

Squire, L. R., *Memory and Brain*, Oxford: Oxford University Press, 1987.

Squire, L. R. and Butters, N. (eds.), *Neuropsychology of Memory*, 2nd edition, New York: Guilford Press, 1992.

Yates, F., *The Art of Memory*, London: Penguin, 1966.

Zeki, S., *A Vision of the Brain*, Oxford: Blackwell Scientific Publications, 1994.

8   Memory and Neural Networks

TERRENCE J. SEJNOWSKI

The brain's operation depends on networks of nerve cells, called neurons, connected with each other by synapses. Scientists can now mimic some of the brain's behaviours with computer-based models of neural networks. One major domain of behaviour to which this approach has been applied is the question of how the brain acquires and maintains new information; that is, what we would call learning and memory. Neural networks employ various learning algorithms, which are recipes for how to change the network to store new information, and this chapter surveys learning algorithms that have been explored over the last decade. A few representative examples are presented here to illustrate the basic types of learning algorithms; the interested reader is encouraged to consult recent books listed in the section on Further reading, which present these algorithms in greater detail and provide a more complete survey.

It is important to keep in mind that the models of neural networks that are simulated in computers are far simpler than the highly complex and often messy neural systems that nature has devised. Although much simpler, neural network models capture some of the most basic features of the brain and may share some general properties that will allow us to understand better the operation of the more complex system. Neural network models are built from simple processing units that work together in parallel, each concerned with only a small piece of information. The representation of an object, for example, is typically represented as a pattern of activity over many of these processing units, and learning to recognize a new object occurs by changing the connection strengths or weights on the links between units. Long-term memories are therefore stored in the neural network, resulting in a close relationship between how information is represented, processed, stored and recalled.

If there is no external supervision, learning in a neural network is said to be unsupervised; when there is an external 'teacher', the learning is supervised. If the teacher provides only a scalar feedback (a single number indicating how well the network has performed), it is called reinforcement learning. A distinction should also be made between a learning *mechanism* and a learning *algorithm*. For example, a Hebbian synapse (named after the Canadian neuropsychologist, Donald Hebb) is a learning *mechanism* that has been found in many parts of the brain, in which the strength of the connection across a neural synapse changes according to the correlation between activity in the presynaptic and post-synaptic neurons. In other words, if neurons fire together frequently, then the synaptic link between them is strengthened. Hebbian synapses are often thought to reflect a particular type of learning algorithm, but it is possible to use a Hebbian mechanism to implement any or all of super-vised, reinforcement and unsupervised learning algorithms.

The time scale for learning can vary over a wide range, and different values of this range are suited to different types of learning. A fast learn-ing rate is good for one-shot memorization, which is appropriate for remembering unique items or events that may only be presented once. Representational learning is slower, since many exemplars from each category need to be compared before the salient distinctions can be extracted that characterize the category. An example of the former is remembering a specific episode involving an elephant, one time when you were at the zoo. What you know about an elephant (what it looks like, where it comes from, how it differs from a rhinoceros or a giraffe), on the other hand, is representational knowledge acquired gradually. Reinforcement learning is typically slow, because the amount of infor-mation gathered on any one trial may be quite small and must be inte-grated over trials before the network can converge on the desired knowl-edge or behaviour.

The processing units used in most neural networks are simplified cari-catures of neurons that lack many features of real neurons in the brain, such as their anatomical structure (e.g. dendrites that form synapses with nearby axons from other neurons, as in Figure 1, *left*). Two

Terrence Sejnowski

Neurons as Processors

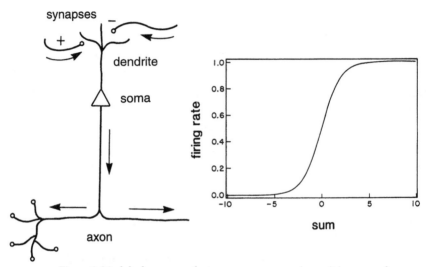

Figure 1 Model of a neuron that serves as a processing unit in a neural network model. *Left*: Schematic model of a processing unit receiving inputs from other processing units on the dendrites, and sending information out along the axon. Neurons are connected together through synapses that can be excitatory (positive weight) or inhibitory (negative weight). *Right*: Input–output function for a processing unit. The total synaptic input at a given moment in time is transformed by the sigmoidal function into an output firing rate.

essential features of real brain systems have, however, been retained. First, the links or connections between processing units in an artificial neural network are a direct analogy of synapses between neurons, whose weighted activity is often linearly summed. Secondly, the non-linear output function that communicates the integrated activity to other network units, typically through a threshold function (Figure 1, *right*), is also based on our understanding of the interaction of real neurons. These simplified models have the virtue of supporting rigorous mathematical analysis that is essential for achieving a deeper understanding of the performance of the network. Simple network models are also the starting point for more sophisticated models that more accurately reflect further properties of real neurons.

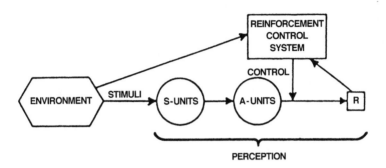

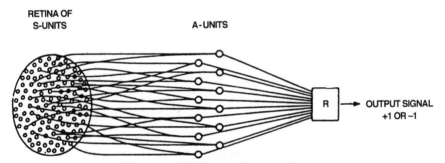

Figure 2 Most of the essential properties of modern multilayer 'perceptrons' can already be found in the earliest versions. *Top*: Learning control system used for training a perceptron. The environment provides sensory stimuli that impinge on the sensory receptors (S-units) and a teaching signal that is used to change the strengths of the connections to the single output unit (R). *Bottom*: There is a fixed set of connections between the sensory receptors and the association units (A-units) and a set of variable connections from the A-units to the output unit. The goal of the perceptron was to classify all sensory patterns into two classes, those that belong to a category, such as the numeral '2', and those that do not.

Supervised learning

A simple learning procedure exists for automatically determining the weights in a feedforward network (i.e. one in which activity flows from input to output without feedback loops) having a single layer of variable connection strengths between the input and output layers (Figure 2). It is an incremental learning procedure that requires a teacher to provide the network with correct outputs for a set of input patterns; with each input pattern, the weights in the network are slightly altered to improve its performance. If a set of weights exists that can solve the problem, then a convergence theorem guarantees that such a set of weights will be found; this has been demonstrated for the simple two-layer feedforward networks of binary (two-valued) units known as 'perceptrons' (Figure 2), introduced by Frank Rosenblatt in his book on *Principles of Neurodynamics* (1959), and for networks with continuous-valued units as shown by Bernard Widrow and Ted Hoff who introduced the least-mean-squared (LMS) algorithm.

These learning procedures are error-correcting in the sense that only information about the discrepancy between the desired output provided by the teacher and the actual output given by the network is used to update the weights. A serious limitation of a feedforward network with one layer of modifiable weights is that only simple problems can be learned. Fortunately, learning algorithms have recently been discovered for more complex networks that can support non-linear mappings, like those commonly found in neural processing. In 1983, Geoffrey Hinton (a computer scientist and psychologist at University College London) and I generalized the perceptron learning algorithm to a multilayer network with feedback connections, called the Boltzmann machine. Shortly thereafter, David Rumelhart, a psychologist at Stanford University, together with Geoffrey Hinton and Ronald Williams, generalized the LMS learning rule to a multilayered feedforward architecture by the backpropagation of errors.

In the most typical case of a multilayered network, there is an extra layer of 'hidden' units, so called because they do not correspond directly to either input or output but rather learn a non-linear mapping between them; the network therefore has two sets of weights rather than one. A concrete example of a feedforward network with hidden units is illustrated in the next section. In error backpropagation, the error (discrepancy between the correct and actual response of the network) is first used to adjust the weights on connections from the hidden layer to the output layer. Next, the error contribution is backpropagated by the chain rule of differential calculus to the hidden layer, where it can be used to adjust the weights on connections from the input units. Although backpropagation has not been demonstrated in the brain, it is perhaps the best-known network learning algorithm and has been used to solve many problems of practical and/or theoretical interest. Such networks are nearly impossible to design by explicit computer programming. By examining successful networks that have learned by backpropagation, we have discovered new ways of thinking about distributed representations, in which each unit or neuron represents a tiny part of the solution. This is illustrated by NETtalk, a network that was trained by Charles Rosenberg and me to pronounce English text in 1985.

Lessons from NETtalk

English orthography is characterized by a strikingly inconsistent relationship between the spellings and the pronunciations of words. For example, the *a* in almost all words ending in *-ave*, such as *brave* and *gave*, is a long vowel, but *have* deviates from this typical spelling-sound pattern. There are also words such as *read* or *bass* whose pronunciation can vary with their grammatical role or contextual meaning. NETtalk demonstrates that even a small single network can learn to capture most of the regularities in English pronunciation while at the same time absorbing many of the exceptions. The feedforward architecture of this network is shown in Figure 3. First, a sequence of seven letters in an English text was used to activate units on the input layer, then the states of the hidden units were determined by activity feeding forward from the input to the hidden layer, and, finally, the states of the phoneme units at the top were calculated by converging inputs from the hidden layer. The hidden units neither received direct input nor had direct output, but were used by the network to form internal representations that were appropriate for solving the problem of mapping letters to phonemes. The goal of the backpropagation learning algorithm was to search the space of all possible weights on connections in the network for a set that performed the mapping with a minimum of error.

The network was trained on a subset of the words from a 20 012 word dictionary, and new words were used to test its ability to generalize what it had learned. Learning proceeded one letter at a time. Early in the training period, the network learned to distinguish between vowels and consonants; however, it tended to translate all vowel letters to the same

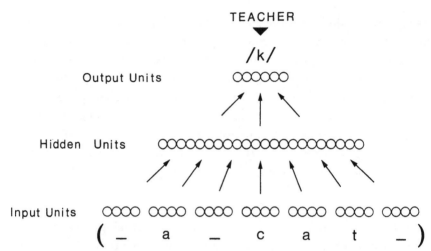

TEACHER

/k/

Output Units

Hidden Units

Input Units

(_ a _ c a t _)

Figure 3 Schematic drawing of the NETtalk network architecture. A window of letters in an English text is fed to seven groups of input units. Information from these units is transformed by an intermediate layer of 80 'hidden' units, which project to the output layer corresponding to the sound of the centre letter. There was a total of 309 units and 18 629 weights in the network. In this example, the 'teacher' instructs the output that the correct phoneme for the letter 'c' in 'cat' is the 'hard c' sound /k/.

vowel sound and all consonant letters to the same consonant sound, which resulted in a babbling effect. A second stage occurred when word boundaries were recognized, and the output then resembled pseudo-words. After just a few passes through the dictionary, most words were intelligible.

By studying the patterns of activation amongst the hidden units, it was possible to understand some of the coding methods that the network had discovered. On average, for any given input, about 20% of the hidden units were highly activated while most of the remaining hidden units had little or no activation. Thus, the coding scheme could be described as neither a local representation, which would have activated only one or two hidden units specific to a given input, nor a 'holographic' representation in which all hidden units would have participated to some extent. When the hidden units were analysed with statistical techniques, it became apparent that the most important distinction was the complete separation of consonants and vowels. However, within each of these two groups the clustering had a different pattern. For the vowels, the next most important variable was the individual vowel letter (*a, e, i,* etc.), whereas consonants were clustered according to the similarity of their sounds.

It was surprising to find that, when NETtalk was trained on new words, it displayed a phenomenon called the 'spacing effect' in human learning. In 1885, Hermann Ebbinghaus, the pioneering German psychologist, in his book *Memory: A Contribution to Experimental Psychology*, noted that 'with any considerable number of repetitions, a suitable distribution of them over a space of time is decidedly more advantageous than the massing of them at a single time'. The spacing effect has been found to apply to learning across a wide range of stimulus materials and tasks, semantic as well as perceptual/motor, and has even been found when the repetitions of information occur across different modalities (such as auditory and visual presentation) or across different languages. The ubiquity of this phenomenon suggests that spacing reflects something of central importance in learning and memory. However, despite over 100 years of research, there is no adequate, or at least simple, explanation for the spacing effect. In the case of NETtalk, the spacing effect arose because new words needed to be integrated into the representation of the previously learned words in a distributed and incremental way that would not interfere with previous learning, and this provides a hypothesis about the nature of the spacing effect in human learning.

NETtalk is an illustration in miniature of many aspects of learning. First, the network starts out with considerable 'innate' knowledge in the form of input and output representations that were chosen by the experimenters, but with no knowledge specific for English – the network could have been trained on any language with the same set of letters and phonemes. Secondly, the network acquired its competence through practice, went through several distinct stages, and finally reached a significant level of correct performance. Thirdly, and very important as a simulation of real memory in the brain, the learned information was distributed throughout the network such that no single unit or link was essential. If the trained network were subsequently damaged, in an analogue to brain damage from stroke or neurodegenerative conditions like Alzheimer's disease, then the network behaved in a fashion that has come to be known as *graceful degradation*. That is, its responses became 'noisier' or more error prone but the network's performance did not simply collapse; in particular, correct responses to specific inputs (words) were not suddenly lost. Finally, the effect of temporal ordering and spacing during learning of new information was remarkably similar to that in humans.

Despite these similarities to human learning and memory, NETtalk is too simple to serve as a good model for the acquisition of reading skills in humans. The network attempts to accomplish in one phase what occurs in two major phases of human development. Children learn to talk first; only after representations for spoken words and their meanings are well developed does the child learn to read. It is also possible that, in addition to our ability to use context-dependent letter-to-sound correspondences, we have access to articulatory representations for whole words, but there are no whole-word level representations in the network. It is perhaps surprising that the network was capable of reaching a significant level of performance using a window of only seven letters. Despite these limitations, we learned from NETtalk and many similar examples that a relatively simple learning algorithm is capable of acquiring much more complex behaviour than we had previously imagined. This gave us confidence that other learning algorithms could be found that were more biologically plausible and even more powerful. The algorithms discussed below describe several staging posts along the way towards this goal.

Reinforcement learning

The external teacher in a supervised learning algorithm should not be taken too literally, since one brain area could serve as teacher and provide the information needed to train another brain area. One mechanism for this might be to have a sensory system 'teach' the motor system to mimic a sensory input. For example, when songbirds learn to sing, there is an auditory phase during which the young bird listens to and becomes familiar with the tutor song of an adult, followed by a sensorimotor phase of successive approximation and improvement leading to adult song. It is believed that a 'template' of the tutor song is formed in the auditory system during the first phase and that, during the sub-

sequent sensorimotor learning, the developing song of the bird is compared to the stored template, and changes are made to the motor system
to improve the song.

There are specialized circuits in the bird's brain that are used to
process birdsong. An 'anterior forebrain pathway' is needed for song
learning but is not required for adult performance after learning is
completed. Thus, in this form of mimicry, the 'teacher' may be a sensory
representation stored in one part of the brain that is used as a reference
to train the motor system through incremental learning. Recently, Kenji
Doya and I developed a model of birdsong learning that uses a form of
supervised learning called reinforcement learning. First, small changes
were made to the connection strengths that control the song production.
The degree of match between the song produced and the stored auditory
template was then used to determine whether performance had
improved, and whether to maintain or further refine the changes made
to the connection strengths.

Often, only a crude feedback signal is available from the external
world after a sequence of behaviours. In classical conditioning, a well-
studied form of reinforcement learning that is found in a wide range of
species, an animal may have a particular sequence of behaviours
strengthened by finding a food reward, or weakened if the behaviour
results in an aversive experience. In psychology, conditioning has been
studied mainly through observing the behaviour of animals such as rats
or birds, as summarized by Nick Mackintosh of Cambridge University in
his book *The Psychology of Animal Learning*. Recently, however, it has
been possible to follow the sensory and motor pathways inside the brain
that are responsible for this form of learning. Classical conditioning is
regulated by diffuse ascending neurotransmitter systems, arising from a
relatively small number of neurons in the brainstem, that project
throughout large regions of the forebrain. Similar diffuse neurotransmitter systems are also found in invertebrate animals, as illustrated
next for the bee.

Bee foraging and temporal difference learning

Bees are among the most intelligent insects and survive as a colony by
foraging for nectar in plants that may be at some distance from the hive.
In particular, Randolf Menzel at the Free University in Berlin has shown

that honeybees can learn to associate a sucrose reward with an odour using proboscis extension as a behavioural assay for learning. The presence of sugar in nectar will automatically produce a proboscis extension in the bee. When paired with sucrose, an odour that previously had no effect on proboscis extension reliably elicits this response; this is straightforward classical conditioning. Martin Hammer, also in Berlin, using a technique of intracellular recording, has identified a single neuron in honeybees (called VUMmx1) that may mediate this simple form of reinforcement learning. Hammer's new discovery was that the nerve-cell firing of VUMmx1 by injection of electrical current can substitute for sucrose in a conditioning experiment: after pairing between VUMmx1 activation and an odour, the odour by itself elicits proboscis extension. Thus the activity of VUMmx1 can substitute for a reinforcing stimulus like nectar. In fact, direct application of the neurotransmitter released by VUMmx1 can also substitute for a rewarding stimulus.

Together with Read Montague, at the Baylor College of Medicine in Houston, and Peter Dayan, at University College London, I recently modelled the foraging behaviour of bees using a reinforcement learning algorithm (Figure 4). The model is based on the principle of prediction by temporal differences introduced by Richard Sutton, a psychologist, and Andrew Barto, a computer scientist. There is a central role in the model for VUMmx1, which is responsible for teaching the bee to predict the future reward consequent on sensory stimuli. As indicated above, if an odour temporally precedes the delivery of nectar, then, through a predictive form of the Hebbian learning algorithm, the inputs to VUMmx1 are strengthened. This learning algorithm is Hebbian because it depends on the conjunction of pre- and postsynaptic activity; however, the postsynaptic activity is not the summed synaptic input but the *prediction* error, calculated as the difference between the actual unconditioned reward stimulus (nectar) and the amount of reward expected on the basis of changing predictions about the future. This model accounts for data obtained from a large number of behavioural experiments on bee learning, including risk aversion behaviour.

Wolfram Schultz, a neurophysiologist at the University of Fribourg, Switzerland, has recorded from single neurons in primates and found firing patterns in response to rewarding stimuli that are analogous to

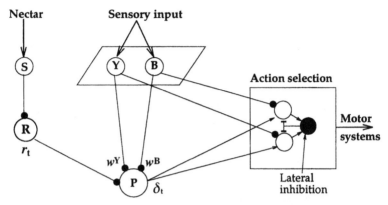

Figure 4 Neural architecture for a model of bee foraging. Unit P learns to makes predictions about future expected reinforcement. Sensory input (units B and Y) are activated by blue and yellow flowers. S is activated while the bee sips the nectar. Learning takes place at the weights w after comparing expected reward with the actual reward r_t originating from unit R. During foraging, the sensory impression of a flower evokes a predicted reward δ_t from P, which is used by the action selection network to decide whether to approach or to avoid the flower. In the decision network, lateral inhibition is used to ensure that only one action is selected between the two possibilities, a form of winner-take-all circuit.

those observed in VUMmx1 in bees. Early in learning, these neurons fired reliably to reward, but later in learning they no longer responded directly to the reward; instead they were activated by the earlier sensory stimuli that reliably predicted the reward. This is exactly what would be produced by a predictive Hebbian learning rule that modifies the inputs to these neurons, and underlies the model's explanation for phenomena of classical conditioning. The output from such a cell may therefore carry information about errors in the prediction of future rewards consequent on specific stimuli. This interpretation was supported in an experiment in which the expected reward was withheld: neurons that no longer responded directly to the reward showed a *decrease* in activity when the reward should have occurred. The predictive principle of temporal difference learning provides computational explanations for many otherwise puzzling facts about learning in the brain.

Predictive Hebbian learning may be a universal mechanism that is important for orienting animals to stimuli that lead to future reward.

The brain is probably organized to make predictions about the importance of sensory stimuli for survival in an uncertain world, and to use these predictions as a basis for appropriate action. In the example given here, only a single action was trained; but temporal difference learning can be used to learn a sequence of actions leading to a delayed reward. For example, Gerald Tesauro at the IBM T. J. Watson Laboratories in Yorktown Heights, New York, has used temporal difference learning to train a network to play backgammon at a level that is competitive with the best human players. The network improved from random moves to master level by playing against itself, and the only reward that it received was the outcome of each game – won or lost. This is a remarkable achievement that illustrates the potential for temporal difference learning to master complex tasks.

Unsupervised learning

When there is no explicit teacher, there is still much to learn. As the former New York Yankee baseball player Yogi Berra once said, 'You can observe a lot just by watching'. The world is information rich, and the problem of the learner is to extract from the constant stream of sensory information a summary of the regularities in this information that will help to improve the recognition of familiar items or events and the detection of new ones. One good strategy for unsupervised learning is to try to predict future sensory states as a basis for developing predictive representations. This type of representational learning is not guided by a specific task, but may none the less be biased in directions that are either innate or influenced by experience. People who are amnesic following damage to the hippocampal formation (see Wilson, Chapter 6) are profoundly impaired on explicit learning, but have significantly preserved ability to learn tasks that might be based on implicit memory at an early representational level. The development of the nervous system also offers important clues to unsupervised learning, since many of the same mechanisms that organize neural interactions during development are also used, in modified forms, in adult brains. In particular, there are several diffuse systems of neuromodulators ascending from the brainstem that support neural plasticity during development.

An example of an unsupervised learning algorithm that can be

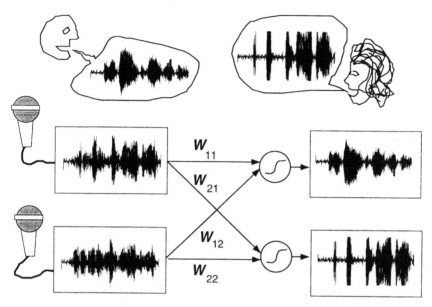

Figure 5 Illustration of the blind separation problem. Two microphones record signals from two people speaking simultaneously, but with different amplitudes at each microphone. The task is to find a set of weights w_{ij} that invert the linear mixtures on the inputs to produce outputs that are the original independent signals from each speaker. No specific information is given about the signals. The independent component analysis algorithm can solve this problem.

implemented by a neural network is a form of statistical analysis closely related to Hebbian synaptic plasticity known as principal component analysis (PCA). A new class of unsupervised algorithms, called independent component analysis (ICA), can extract higher-order information, and has been the subject of intense investigation by a group of researchers worldwide. Anthony Bell and I have proposed a framework for 'blind' source segregation, based on an idea introduced by Simon Laughlin at Cambridge University in the context of visual information transfer, that can separate signals into their independent components, as illustrated in Figure 5. The ICA learning algorithm we derived has proven effective in analysing many types of data, including natural images, natural sounds, and electroencephalographic data in which the independent sources correspond to electrical activity in different regions of the brain. One result from analysing visual images in this framework will be summarized here.

Edge detectors in visual cortex

The classic experiments of David Hubel and Torsten Wiesel on neurons in the visual cortex showed that many cells preferred specific features, such as edges and lines of particular orientation in a small part of the visual field, different for each neuron, called its receptive field. This led Horace Barlow at Cambridge University to search for coding principles that would predict the formation of such feature detectors. Barlow suggested that edges are suspicious coincidences that arise at the end of a redundancy reduction process, the purpose of which is to make the activation of each feature detector as statistically *independent* of all others as possible.

Anthony Bell and I recently applied our ICA learning algorithm to images of natural scenes including trees, foliage and bushes. The training set consisted of 17 595 input patterns consisting of 12 pixel × 12 pixel patches from the images. The filters and basis functions resulting from training on natural scenes are displayed in Figure 6. The general result is that the filters found by ICA learning are localized and mostly oriented. These filters produce output distributions of very high sparseness; that is, when an image is convolved with all of the ICA filters, only a few of them are highly activated and most of them have nearly zero output. The general properties of the ICA filters, which were designed to be as statistically independent as possible, are similar to those of oriented simple cells in primary visual cortex, supporting Barlow's intuition regarding the independence of cortical feature-detector cells.

ICA is a powerful unsupervised learning algorithm, but it is limited to one layer of processing units, and the output is a linear transformation of the inputs. The architecture of cerebral cortex is organized in a hierarchy of processing layers, starting with the primary sensory areas and ascending to more complex and eventually multimodal representations. At the top of the hierarchy, neural activity from all the sensory modalities (vision, hearing, touch, balance, etc.) funnels into the hippocampus, a part of the brain thought to be necessary for creating long-term mem-

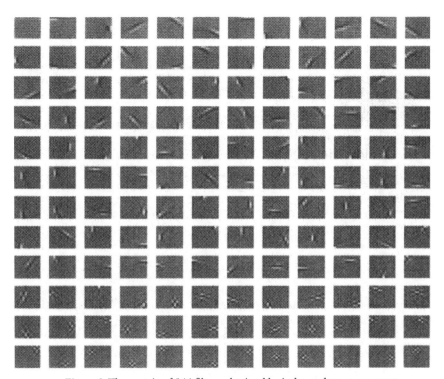

Figure 6 The matrix of 144 filters obtained by independent component analysis training. Each filter is a 12 × 12 patch that produces an output depending on how well the excitatory regions of the patch (white) match the high intensity regions of the image and avoid the inhibitory parts of the filter (black). Most of the filters have a preferred orientation, which resemble the response properties of simple cells in visual cortex, though some are localized chequerboard patterns that produce responses in two perpendicular directions. Only a few of these filters will be highly activated by any given image, which produces a sparse encoding of the image by the filters. These filters form a complete set in the sense that any patch of image can be exactly reconstructed knowing only the activity levels of each filter.

ories of specific objects and events. Damage to the hippocampus (which is common, for example, in Alzheimer's disease) interferes with new and recent learning; but if the hippocampal damage occurs some substantial time after a learned experience, then there is little or no significant disruption of the memory. This implies that, although the hippocampus mediates consolidation of memory, the neural patterns corresponding to

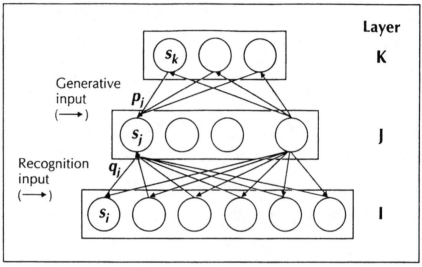

Figure 7 Wake–sleep network model. Sensory inputs (s_i) in layer I originate in the bottom layer and feedforward connections (q_j) carry this information through a layer J of hidden units (s_j) to the output units (s_k) top layer K in the recognition or wake phase. The feedback connections (p_i) from top to bottom provide a generative input to the bottom layer during the sleep phase.

long-lasting memories are stored in other brain regions. Unsupervised learning algorithms are needed that can create a hierarchy of sensory representations similar to those found in the brain.

Wake–sleep learning

Geoffrey Hinton, Peter Dayan and their colleagues have recently presented a powerful new theoretical framework for creating efficient memory representations in hierarchical feedforward neural networks (Figure 7). The framework suggests computational explanations for why we may need to sleep and why the memory consolidation process can take so long. In this model, when the external inputs to the network have been suppressed (just as external stimulus inputs to the brain are suppressed when we are asleep), then the feedback connections generate patterns on the input layers of the network that correspond to representations at the higher level. During this generative 'sleep' stage, the strengths of the feedforward synaptic connections are altered. Conversely, during the 'awake' state, the sensory inputs drive the feedforward system; in this phase, weights on the feedback connections can be modified. This two-phase process produces an internal, hierarchical representation of an experience that is economical and able to generate typical input patterns. The learning mechanisms are biologically possible since, unlike supervised learning algorithms such as backpropagation that require a detailed teacher, wake–sleep learning depends only on locally available signals and there is no external teacher.

The wake–sleep learning algorithm attempts to capture the regularities of sensory inputs with an internal code that is componential: it is capable of representing not just whole individual items but components and features that are common to many objects. Because these statistical components are not apparent without comparing many sensory experiences, the training process is gradual, in the sense that only small changes are made during any one wake–sleep cycle. From this point of view, the hippocampus could serve as the highest level of representation. Through a generative process involving feedback connections from the hippocampus to the cortex, feedforward connections in the cortex that are responsible for the primary representations of sensory information are modified. The forward recognition network and the feedback generative networks are the mutual statistical inverses of each other.

The wake–sleep model makes testable predictions for the function of feedback connections and the times at which they should be modifiable. It has proved difficult to assign any function for cortical feedback connections, and the wake–sleep model explains why this might be. Experiments designed to test the effects of these connections have been performed during the awake state; but according to the model, they should drive physiological activity only during sleep. Also, the model predicts that the feedback and feedforward projections should be modifiable at different times: feedforward connections during sleep, and feedback connections during the awake state. The types of experiment that are now possible using multielectrode recordings should allow these predictions to be tested directly.

Although the links between sleep and memory are intriguing, the different phases of wake–sleep learning may not correspond in such a straightforward fashion to awake or sleeping states. There are several different phases of sleep, one of them being dream, or rapid eye-movement (REM), sleep, so it is not clear which sleep state should be associated with the 'sleep' phase of the wake–sleep learning algorithm. It is also possible that both phases alternate even during states of consciousness, with the sleep phase of the algorithm corresponding to lapses of attention and daydreaming. None the less, it is worthwhile to explore the possibility that the same neural network may have quite different properties in different states, including differences in the forms of neural plasticity. The state of the cortex is influenced by neuromodulators, which are known to affect neural plasticity, so mechanisms are already known that could implement these state changes.

Biological and mathematical foundations

Many different types of memory systems have been found in different parts of the brain. Even the retina can adapt to light intensities over a wide range of temporal and spatial scales, a form of short-term sensory memory. Therefore we should expect not a single, universal model of memory but rather a diversity of memory models, and this is reflected in a diversity of learning algorithms. In this chapter, only a few, representative examples of neural network models of learning and memory were presented to illustrate this richness.

The focus has been on learning algorithms in which the parameters that represent long-term changes in memory are the strengths of the synapses. There are other variables and parameters in network models that may also be important, including the number of units (neurons) in the model, the short-term dynamics of connections (synapses), and the properties of the non-linear input–output function. For example, it is possible for a dynamical network model with feedback connections to exhibit long-term memory without changing any of the weights on connections, by changing only the time courses of the synaptic activities.

A memory system can be considered a form of information compression for an ensemble of sensory inputs or sensorimotor transformations. Thus each memory system in the brain can be analysed to discover how well it is able to perform information compression, and what types of information are most efficiently represented. A mathematical principle, called Minimum Description Length (MDL), has been helpful in developing new learning algorithms and provides a useful framework for organizing existing algorithms. The MDL framework may also help us to understand changes that occur in the developing brain.

One of the most exciting aspects of neural network research is the extent to which it crosses disciplines, including neuroscientists interested in understanding how brain systems are organized, cognitive scientists interested in the properties of minds, and engineers who want to mimic some of the brain's impressive capabilities. We are just beginning to understand learning and memory in the simplest neural network systems, and to appreciate the complexity of the many memory systems found in brains.

FURTHER READING

Arbib, M. A., *The Handbook of Brain Theory and Neural Networks*, Cambridge, MA: MIT Press, 1995.

Ballard, D. H., *An Introduction to Natural Computation*, Cambridge, M.A.: MIT Press, 1997.

Bishop, C. M., *Neural Networks for Pattern Recognition*, New York: Oxford University Press, 1995.

Hammer, M. and Menzel, R., 'Learning and memory in the honeybee', *Journal of Neuroscience* **15** (1995), 1617–1630.

Hebb, D. O., *The Organization of Behavior*, New York: Wiley, 1949.

Mackintosh, N. J., *Conditioning and Associative Learning*, Oxford: Oxford University Press, 1983.

Montague, P. R. and Sejnowski, T., 'The predictive brain: temporal coincidence and temporal order in synaptic learning mechanisms', *Learning and Memory* **1** (1994), 1–33.

Quartz, S. and Sejnowski, T. J., 'The neural basis of cognitive development: a constructivist manifesto', *Behavioral and Brain Sciences*, **20** (1997), 537–96.

Tesauro, G., 'Temporal difference learning and TD-Gammon', *Communications of the ACM* **38** (1995), 58–68.

Notes on Contributors

A. S. Byatt taught at University College London before becoming a full-time writer. She has written six novels and four short story collections, and won the Booker Prize in 1990 for *Possession*. Her critical works include two collections of essays: *Unruly Times* (1989), on Wordsworth and Coleridge; and *Passions of the Mind* (1991). She is also author of *Degrees of Freedom* (1965), a study of the fiction of Iris Murdoch, and, with Ignes Sodre, *Imagining Characters: Six Conversations about Women Writers* (1995).

Patricia Fara qualified initially as a physicist and went on to run a small but international company producing audiovisual material on computers and statistics. She now lectures in the history of science at Cambridge University and has been awarded research fellowships at Darwin College and the Australian National University. Her books include *Computers: How they Work and What they Do* (1982), *Sympathetic Attractions: Magnetic Practices, Beliefs, and Symbolism in Eighteenth-Century England* (1996) and *The Changing World* (Darwin Lecture Series, 1996). She is currently writing a book on Isaac Newton and changing concepts of genius.

Jack Goody is a social anthropologist who is at present a Fellow of St John's College, Cambridge. He has carried out fieldwork in Ghana and has done theoretical studies on literacy (for example, *The Domestication of the Savage Mind*, 1977), on kinship and family, and on comparative aspects of cultural history (for example, *Cuisine, Cooking and Class*, 1982). Some of his enquiry has taken the form of contrasting Africa with the major societies of Eurasia (and of drawing attention to similarities between the Orient and the Occident). His latest book, *Representations and Contradictions: Ambivalence about Images, Theatre, Fiction, Relics and Sexuality* attempts to explain opposition to representation found in Europe, the Near East, Asia and to some extent in the oral cultures of Africa.

CATHERINE HALL is a professor in the Department of Sociology at the University of Essex. From September 1998 she will be Professor of Modern British Social and Cultural History at University College London. She is co-author, with Leonore Davidoff, of *Family Fortunes: Men and Women of the English Middle Class 1780–1850* (1987), which explores the relation between gender and class. Her collection of essays *White, Male and Middle Class: Explorations in Feminism and History* (1992) includes some of her more recent work on Englishness and Empire. She is currently completing a book on *Civilizing Subjects: 'Race', Nation and Empire in the English Imagination.*

JULIET MITCHELL is a Member of the International Psychoanalytic Association. She is a lecturer in Gender and Society in the Faculty of Social and Political Sciences, Cambridge University, and a Fellow of Jesus College, Cambridge. She is also an A. D. White Professor-at-large at Cornell University, USA. She recently published (with Ann Oakley) *Who's Afraid of Feminism?* (1997) and is pursuing (with Jack Goody) research on changing family patterns, and writing a re-evaluation of hysteria from a psychoanalytical, social history and feminist perspective. Her books include *Women: the Longest Revolution* (1984), *Psychoanalysis and Feminism* (1974) and *Women's Estate* (1972).

KARALYN PATTERSON is a neuropsychologist in the Medical Research Council's Cognition and Brain Sciences Unit in Cambridge, studying language and memory disorders in adults with various forms of brain disease. She has also spent short periods as a visiting researcher in both Tokyo and Sydney. She is author or co-author of nearly 100 articles in journals and books concerned with cognitive psychology and has served on the editorial boards of five mainstream journals in the field. She was President of the British Neuropsychological Society from 1992 to 1994.

STEVEN ROSE has been Professor of Biology and Director of the Brain and Behaviour Research Group at the Open University since its inception in 1969. He is author of a number of books, including *The Chemistry of Life*, *The Conscious Brain*, *Not in our Genes* and his latest, *Lifelines* (1997). His book *The Making of Memory* (1992) won the Rhône-Poulenc Science Book Prize.

TERRENCE SEJNOWSKI is an Investigator with the Howard Hughes Medical Institute and a Professor at The Salk Institute for Biological Studies. He is also Professor of Biology at the University of California, San Diego and is Part-time Visiting Professor at the California Institute of Technology. The long-range goal of Dr Sejnowski's research is to build linking principles from brain to behaviour using computational models. He has co-authored *The Computational Brain* (1992) with P. S. Churchland.

RICHARD SENNETT is Professor of Humanities at New York University. He is internationally acclaimed for his books on urban history and social criticism, in particular the relationship between the individual and society. His work *The Fall of Public Man* (1977) is now regarded as a contemporary classic and since then he has published several books examining the question of individual freedom (for example, *The Uses of Disorder, Authority* (1993)), three novels and most recently an exploration of people's experience of their bodies as reflected in the form of urban spaces – *Flesh and Stone: The Body and the City in Western Civilization* (1994).

BARBARA WILSON is a Senior Scientist with the Medical Research Council's Cognition and Brain Sciences Unit in Cambridge. Prior to this she worked at the Rivermead Rehabilitation Centre in Oxford and the Department of Rehabilitation at the University of Southampton Medical School. She has published many articles, mostly on rehabilitation of cognitively impaired brain-injured adults, and her books include *Rehabilitation of Memory* (1987), *Clinical Management of Memory Problems* (1992) and the *Handbook of Memory Disorders* (1995). She has lectured throughout Europe, America and Australia and is Editor-in-Chief of the journal *Neuropsychological Rehabilitation*.

Acknowledgements

INTRODUCTION

Figure 1: From Cowling, M., *The Artist as Anthropologist: The Representation of Type and Character in Victorian Art*, Cambridge: Cambridge University Press, 1989.

CHAPTER 1

Figures 1 and 2: Courtesy of Dr W. Pullan, Darwin College, Cambridge.

CHAPTER 2

The author thanks Margaret Rustin for referring her to the essay by John Steiner, and Gail Lewis for discussions of the workings of the 'blind eye' in relation to 'race'.

Figure 1: From Stedman, J., *Narrative of a Five Years' Expedition, Against the Revolted Negroes of Surinam, in Guiana, on the Wild Coast of South America*, London: J. Johnson, 1796.

CHAPTER 3

Extracts from *Remembrance of Times Past: Swann's Way* by Marcel Proust, translated by C. K. Scott Montcrieff, revised by Terry Kilmartin and D. J. Enright, are published with permission of the Estate of Marcel Proust and Chatto & Windus.

Figure 1: By courtesy of the National Portrait Galley, London.
Figure 2: From Yates, F., *The Art of Memory*, London: Penguin, 1966.

CHAPTER 4

Figure 1: From Meyerowitz, E. L. R., *The Sacred State of the Akan*, Plate 34, London: Faber & Faber, 1951.

Figure 2: From Rattray, R. S., *Ashanti Law and Constitution*, p. 132, Oxford: Oxford University Press, 1929, by permission of Oxford University Press.
Figure 3: From Phillips, T. (ed.), *Africa: The Art of a Continent*, p. 285, London: Royal Academy of the Arts, 1995.
Figures 4, 5 and 6: From Dewdney, S., *The Sacred Scrolls of the Southern Ojibway*, Figures 2, 36 and 35, respectively, Toronto: University of Toronto Press, 1975.

CHAPTER 5

Figure 1: Courtesy of Freud Museum, London.

CHAPTER 6

Figure 1: From Kapur, N., Barker, S., Burrows, E. H., Ellison, D., Brice, J., Illis, L. S. *et al.*, 'Herpes-simplex encephalitis: long-term magnetic resonance imaging', *Journal of Neurology, Neurosurgery and Psychiatry* 57 (1994), 1134–42. Courtesy of BMJ Publishing Group, London.

CHAPTER 7

Figure 1: From Yates, F., *The Art of Memory*, London, Penguin, 1966.
Figure 2: After Larry Squire.
Figure 3: Derived from a video by the Open University, *The Human Brain*, for course SD206.
Figure 4: Reprinted with permission from Shallice, T., Fletcher, P., Frith, C. D., Grasby, P., Frackowiak, R. S. J. and Dolan, R. J., 'Brain regions associated with acquisition and retrieval of verbal episodic memory'. *Nature* **368** (1994), 633–5. Copyright 1994 Macmillan Magazines Limited.
Figure 7: After Valverde, F., *Brain Research* 33 (1971), 1–11, Fig. 5. From Rose S. P. R., *The Making of Memory*, London: Bantam Press.

CHAPTER 8

Figure 1: From Sejnowski, T. J. and Rosenberg, C. R., 'Learning and representation in connectionist models'. In *Perspectives in Memory Research* ed. M. S. Gazzaniga, pp. 135–78, Cambridge, MA: MIT Press, by permission of MIT Press.
Figure 2: After Rosenblatt, F., *Principles of Neurodynamics*, New York: Spartan Books, 1959.
Figure 3: From Sejnowski, T. J. and Rosenberg, C. R., 'Parallel networks that learn to pronounce English text', *Complex Systems* 1 (1987), 145–68.

Figure 4: From Churchland, P. S., Ramachandran, V. S. and Sejnowski, T. J., 'A critique of pure vision'. In *Large-scale Neuronal Theories of the Brain*, ed. C. Koch and J. Davis, pp. 23–60. Cambridge, MA: MIT Press, 1994.

Figure 6: From Bell, A. J. and Sejnowski, T. J., 'The "independent components" of natural scenes are edge filters', *Vision Research* 37 (23) (1997), with kind permission from Elsevier Science Ltd, The Boulevard, Langford Lane, Kidlington OX5 1GB, UK.

Figure 7: Sejnowski, T. J., 'Sleep and memory', *Current Biology* 5 (1995), 832–4.

Index

Numbers in *italics* refer to the Figures and Tables.

Index

CPSIA information can be obtained
at www.ICGtesting.com
Printed in the USA
LVHW012031010119
602328LV00007B/54/P